Money, Capital Formation and Economic Growth

Money, Capital Formation and Economic Growth

International Comparison with Time Series Analysis

Masanori Amano
Chiba University, Japan

palgrave
macmillan

First published 2013 by
PALGRAVE MACMILLAN

Palgrave Macmillan in the UK is an imprint of Macmillan Publishers Limited, registered in England, company number 785998, of Houndmills, Basingstoke, Hampshire RG21 6XS.

Palgrave Macmillan in the US is a division of St Martin's Press LLC, 175 Fifth Avenue, New York, NY 10010.

Palgrave Macmillan is the global academic imprint of the above companies and has companies and representatives throughout the world.

Palgrave® and Macmillan® are registered trademarks in the United States, the United Kingdom, Europe and other countries

ISBN: 978–1–137–28182–1

This book is printed on paper suitable for recycling and made from fully managed and sustained forest sources. Logging, pulping and manufacturing processes are expected to conform to the environmental regulations of the country of origin.

A catalogue record for this book is available from the British Library.

A catalog record for this book is available from the Library of Congress.

10 9 8 7 6 5 4 3 2 1
22 21 20 19 18 17 16 15 14 13

Contents

Figures

Tables

Preface and Acknowledgments

This book comprises four variations on a broad theme, namely empirical macro-dynamics. These variations are: (1) the 'missing equations', (2) the Kalman filters on the natural rate of unemployment and Okun's Law, (3) causality relationships between finance and development, and (4) causality from domestic investment or from exports to output growth.

Although the first variation does not belong to time series analysis in the narrow sense of the word, I decided to include it, because the topic has long been regarded as an unresolved question in macro-dynamics, which is developed on a long time horizon and has some bearing on policy discussion.

Variations (1) and (2) can broadly be grouped within Phillips curve analysis. Hence, although these topics are considered on roughly a half century time horizon, their policy implications have to be sought within short- or medium-term horizons, of at most five to ten years.

Variations (3) and (4), on the other hand, are broadly characterized as examinations on the causality among variables changing over time. Also, the policy significance from these topics will have to be dug out on a horizon extending beyond five years. Although I refer to policy meanings of the topics we will deal with, which is reflected in the book's make-up, that is, most chapters are conducting international comparison, I will draw readers' attention to the main intention of this book which places at least as much emphasis on positive analyses as policy considerations of macro-dynamics.

This book has grown out of my past research and teaching experiences dealing with empirical macro-dynamics, and is made up of four parts.

Part I (Chapters 1–3) sets out to formulate and estimate the 'missing equation.' This concept was first used by Milton Friedman and referred to an equation that closes (makes solvable) two simple macro models, a monetarist model where aggregate output is constant, and a Keynesian model where the price level is constant (Friedman 1970; Gordon 1974; see References in Chapter 1).In this book, I expand empirical macro-dynamics to cover intermediate situations, including Friedman's two polar cases, where, simultaneously with nominal income changes, both output and the price level change. In other words, I deal with macro models which are partly monetarist and partly Keynesian. Each chapter in Part I deals with a different group of countries (the USA, the UK, and Japan; some developing countries; and some OECD countries) and empirically examines the intermediate situations between Keynesian and monetarist (or new classical macroeconomic) extreme cases. Part I then estimates the 'missing equations' (the size of eo and ep, nominal income elasticities of output and price level) which measure how output and price level respond to various macroeconomic demand and supply disturbances.

Part II (Chapters 4–5) deals with the NAIRU (non-accelerating-inflation-rate of unemployment, alternatively called the natural rate of unemployment) and Okun's Law. The latter is a trade-off relation between output and the rate of unemployment. Although the NAIRU is thought to be a constant number in the text-book treatment, it actually is not. This part estimates the evolution of NAIRU for post-war USA, UK, and Japan. Okun's Law is also estimated for those three countries, using the Kalman filter method for both concepts, and then the outcomes for the three countries are compared and evaluated.

Parts I and II focus on topics broadly included in the Phillips curve analysis, which has had a permanent seat in monetary policy discussion. Because of instability and analytical vagueness surrounding the Phillips curve, however, it does, and might continue to, occupy the central position in macro and policy analysis. Also, the analysis developed in the first two parts is relevant to the short-run or medium-run time horizon (of at most five to ten years), since monetary policy is commonly thought to be neutral in the long-run horizon.

Part III (Chapters 6–7) is devoted to examining the causal relationship between financial development and economic growth, using VAR (vector autoregression; multi-variable time series analysis) frameworks involving cointegration and error-correction mechanisms. This part of the book is the first attempt in the literature that identifies (that is, pins down) the cointegrated relations using economic theory. This method was first proposed formally by Wickens (1996) and Pesaran and Shin (2002; see References in Chapter 6).

The first half of Part III examines, for the USA, the UK, and Japan, the cause and effect between financial development and real-side development. We have so far seen much similar work using cross-country regression analysis, but the method we use here, within the time series framework, goes beyond conventional analysis by incorporating direct finance (market finance). The latter half of Part III discusses, also for the three countries, whether the proportion between market finance and bank finance affects the country's economic growth, as well as causality relations between financial development and real-side development. In this way the second half of Part III may be viewed as an examination of the relevance of Modigliani–Miller's theorem on a long-term horizon.

Part IV (Chapters 8–9) consists of comparing causality patterns from domestic investment in fixed capital to output growth, and from exports to growth. In other words, we inquire, using a similar method to the one in the Part III above, whether the growth of the country in question was investment-led or export-led. Chapter 8 focuses on prewar and postwar USA, UK, and Japan, while Chapter 9 tries to shed light on 13 countries consisting of mainly emerging economies.

Overall, the book addresses several methods of dealing with empirical macro-dynamics. The book is intended to furnish the readers with detailed descriptions of the methods, as well as show practical examples of new devices in VARs.

As for the relationship between the first two parts and the second two, the former is concerned with short (or medium)-run policy matter, while the latter deals with long-run development policy; in this sense, the first two and the second two are complementary in topics and purposes.

The book presupposes that the readers have at least a basic knowledge of linear algebra, calculus, and econometrics. Being a converter from theory to econometrics (though I still believe in the former), I have tried to be conscious of economic implications behind statistical discussions.

Many colleagues and friends have, in one way or another, influenced the course of my research, resulting in this book. An initial impetus for time series analysis came from the late Clive Granger, who visited the University of Alabama, Tuscaloosa, when I was staying there as an exchange faculty-member. Although our conversation did not extend to a technical one, it was a pleasant and memorable stimulus for me to spend much of my time on the subject since then.

I would also like to thank Charles Goodhart, then professor at the London School of Economics, and Juro Teranishi, professor emeritus at Hitotsubashi University. As the specialists in money and monetary history of England and Japan, respectively, the two professors were kind enough to be my discussion partners. My thanks should also go to Professors Koji Aoyama and Haruka Yanagisawa, of the Universities of Chiba and Tokyo, respectively, who offered me some technical and other kinds of advice on finishing the manuscript.

Last but not least, I wish to acknowledge the valuable comments and criticisms of two anonymous reviewers; these certainly improved the initial manuscript. Needless to say, none of the people mentioned above is responsible for the content of my work.

Abstracts

Chapter 1 The 'Missing Equations' for Postwar USA, UK, and Japan (p. 3)

This chapter formulates and estimates the aggregate nominal income elasticity of real output (eo) and, hence, the elasticity of the price level (ep) for the USA, the UK, and Japan. The two elasticities concern how annual changes in nominal GDP are divided into output changes and price changes. Similar concepts were suggested in J.M. Keynes (1936) and M. Friedman (1970, 1971). We start by examining firm's optimal pricing behavior, and then, along with short-run determinants of output growth, we derive and estimate equations for eo ($eo + ep = 1$). Estimation shows different responses of eo in the three economies to various demand and supply shocks as well as to inflation expectations.

Chapter 2 Estimating the 'Missing Equations' for Developing Countries (p. 22)

This chapter concerns the 'missing equations' of nine countries mainly chosen from developing areas. The missing equations were originally named by Friedman (1970) and refer to the equations that close the two macro models he discussed, that is, in the quantity theory it is 'total output = given', while in the Keynesian income-expenditure model it is 'the price level = given'. Here, however, we will associate with it the equation describing the proportion of output change in nominal income change (eo). In other words, we apply the equation to intermediate cases between the quantity theory and income-expenditure theory. We derive the equation determining eo from firms' optimization behavior and inquire what factors are the determinants of eo for nine (mainly) developing countries.

Chapter 3 A Quest for the 'Missing Equations' in OECD Countries (p. 37)

This chapter formulates and estimates the aggregate nominal income elasticity of real output (*eo*), and hence, the same elasticity of the price level (*ep*) for eight OECD countries. The two elasticities concern how annual changes in nominal GDP are divided between output changes and price changes. Similar concepts were suggested in J.M. Keynes' main book (1936) and M. Friedman (*Jour. Polit. Econ.* 1970, 1971). We start by examining firms' optimal price adjustments and longer-run profit maximizing behavior, and then, along with short-run determinants of output growth, we derive and estimate equations for *eo* (*eo* + *ep* =1). Estimation shows similar or different responses of *eo* in the eight economies to various demand and supply shocks as well as to inflation expectations.

Chapter 4 The NAIRU, Potential Output, and the Kalman Filter: A Survey and Method of Estimation (p. 59)

This chapter deals with the estimation and interpretation of the NAIRU (or natural rate of unemployment) and potential output, using the Kalman filtering algorithm. The observation equation, describing the dynamic relationship between observable exogenous variables and the unobservable state variable, is made up of several variants of the expectation-augmented Phillips curve. Preceding the estimation we survey the literature in related fields and, toward the end of the chapter, we mention several points to be considered in future work.

Chapter 5 The NAIRU, Potential Output, and Okun's Law: Postwar USA, UK, and Japan (p. 80)

This chapter estimates the NAIRU and potential output for the postwar US, UK, and Japanese economies using the Kalman filter method. The main device of our framework that enables the estimation results for the three countries to be comparable to each other is that we introduce Blanchard's (1997) version of the Okun Law, where he contrasts output growth in excess of the normal growth rate and changes in the rate of unemployment. In the original Okun relationship, the two variables are the output gap in excess of its potential level and the rate of unemployment in excess of its natural level, respectively. This original version involves unobservable variables, while Blanchard's version is composed only of observable variables. Using this correspondence we choose the two error variances in Kalman filter formulations to estimate the above two latent variables as well as Okun's relations for the three countries.

Chapter 6 Finance and Growth: VARs with Cointegration for the USA, the UK, and Japan (p. 99)

This chapter examines causal directions between financial development and economic growth for prewar and postwar USA, UK, and Japan, where finance includes the direct one represented by equity finance. The method used is what may be called a generalized Granger causality test. We estimate the coefficients of long-run cointegrated relations and shorter-run adjustment processes simultaneously and thus efficiently. The chapter also uses the results shown by Wickens, Pesaran, and Shin, which assert the need for imposing some *a priori* restrictions on cointegration coefficients for identifications. We show that the causal patterns are country- and period-specific. This implies that the 'Patrick hypothesis' is not necessarily relevant.

Chapter 7 Financial Structure and Economic Growth: Evidence from the USA, the UK, and Japan (p. 120)

This chapter examines (i) causality patterns between growth in per capita income and financial development and (ii) whether and how changes in financial structure spur economic growth, for the prewar and postwar periods of the three countries. We set up four-variable VAR systems with cointegration and error-correction mechanisms, and estimate coefficients in the above two mechanisms simultaneously. We use the formal results provided by Wickens, Pesaran, and Shin, which assert the need for imposing some economic theory-based *a priori* restrictions on cointegration coefficients for identification. For question (i) we show that most cases exhibit the causality from the financial to real-side development, while for question (ii) most cases do show certain effects although the causal directions are country-specific.

Chapter 8 Has Growth Been Led by Investment or Exports? Prewar and Postwar US, UK, and Japan (p. 143)

This chapter sets up four-variable VAR systems with cointegration and error-correction mechanisms to explore whether per capita output growth has been led *mainly* by domestic investment or by exports for two periods, prewar and postwar, of the three countries, the USA, the UK, and Japan. We use the formal results of Wickens, Pesaran, and Shin, which maintain the need for imposing some *a priori* restrictions on cointegration coefficients for identification. The main findings are as follows: (1) US prewar growth was caused by exports, while its postwar growth was led by investment. (2) UK's prewar growth was induced by both investment and exports, but its postwar growth was influenced by exports. Finally, (3) Prewar Japanese growth was led by investment, while its postwar counterpart was caused by exports.

Chapter 9 Was It Investment or Exports That Led Economic Growth? 13 Developing Country Experiences (p. 157)

This chapter examines, for 13 developing countries, whether domestic investment or exports led their per capita income growth in the postwar period. We construct four-variable VAR systems involving cointegration and error-correction mechanisms. We also use the results shown by Wickens, Pesaran, and Shin, maintaining the need for imposing some *a priori* restrictions on cointegration coefficients for identification. The main conclusions are that in five countries, growth was led by both of the above demand factors. Also, other five countries had investment, and the remaining three had exports, as their main promoter of economic growth.

Part I

1
The 'Missing Equations' for Postwar USA, UK, and Japan

1.1 Introduction

Searching for factors which determine the proportion of output change in nominal income change, and the proportion of price change in nominal income change, has been regarded as one of the unresolved questions in macro-economics. (See Nobay and Johnson 1977; Gordon 2009, ch 7.) In the papers which were intended to describe monetary theory in the monetarist tradition, M. Friedman (1970, 1971) and Gordon (1974) presented frameworks for monetary analysis which describe the quantity theory and the income-expenditure theory. The two frameworks differ in the last equation, which solves the variables of the systems determinately. The difference between the two frameworks described by Friedman was that one made output (national income) fixed for quantity theory, while the other made the price level fixed for income-expenditure theory.

In his 1971 paper, Friedman also showed a model in which nominal income is an endogenous variable, but the division of nominal income into output and the price level was left unspecified. Originally, Friedman called the last (seventh) equation to close both systems 'the missing equation'. However, in this chapter I would like to define the missing equation as that which describes the proportions of output change and price change in nominal income change because Friedman noted in Gordon (1974) that none of his models (that is, those mentioned above) 'have anything to say about the factors that determine the proportions in which a change in nominal income will, in the short-run, be divided between price change and output change'. (Gordon (1974, p.45)).

In spite of the possible importance of the concept, the theoretical or empirical works dealing with the missing equation have not so far been large in number. The early literature of Laidler (1973) and McCallum (1973) correctly recognized the need for empirical implementation of Friedman's analytical proposal, but both authors seem to have limited the number of determinants (independent variables) of the missing equation.

In this chapter we attempt to derive an equation describing the proportion of annual output change in nominal income change (in other words, nominal income elasticity of total output), based on an optimizing behavior of the firm sector, where the equation is a function of short-term aggregate demand components (including money supply), labor market tightness, technical progress, and price expectations. Then we estimate the equation using the two-stage least squares method with forward-looking rational expectations regarding inflation for three countries, the USA, the UK, and Japan, for the postwar period, and also make some comparisons between those countries.

The following empirical work will reveal some interesting contrasts between the three countries regarding short-run output responses to changes in aggregate demand and supply components, inflation expectations and so on.

The next section (Section 2) sets out firms' optimizing behavior and, combining with it an equation describing short-term economic growth of the economy, derives the missing equation (nominal income elasticity of output) as a function of the variables mentioned above. Section 3 estimates the equation for the period 1951 through 1998 using annual data of the three countries (see Section 3 for the reasons behind this period choice), and then compares the empirical results. Finally, Section 4 presents a summary and concluding remarks.

1.2 Firm Behavior and Nominal Income Elasticity of Output

We start with some definitions of variables and concepts. Let us write nominal income (nominal GDP) in some year as Y. Then, writing P for the price level (GDP deflator) and Q for real output (real GDP) of the same year, one obviously has $Y = P \cdot Q$. Also, let gz be the annual growth rate of z ($z = Y$, P, or Q); the descriptions of variables to appear are gathered after Section 4. Then, from $Y = P \cdot Q$,

$$gY = gP + gQ.$$

Next, define *eo* as the elasticity of output regarding nominal income, and *ep* as the elasticity of the price level regarding nominal income. Then one obtains

$$eo = \frac{\Delta Q}{\Delta Y} \frac{Y}{Q} = \frac{\Delta Q/Q}{\Delta Y/Y}, \, ep = \frac{\Delta P}{\Delta Y} \frac{Y}{P} = \frac{\Delta P/P}{\Delta Y/Y}$$

where Δ is a difference operator. If one regards ΔY_t as $Y_t - Y_{t-1}$, where the subscript t refers to some year, then *eo* and *ep* can be written as

$$eo = \frac{gQ}{gP + gQ}, \, ep = \frac{gP}{gP + gQ}. \tag{1.1}$$

Evidently, one has $eo + ep = 1$. Hence the rest of this chapter will focus on the determinants of gP and gQ, to investigate finally on what variables *eo* in each year depends. Annual changes in *eo* for the USA, the UK, and Japan are shown in Figure 1.1.

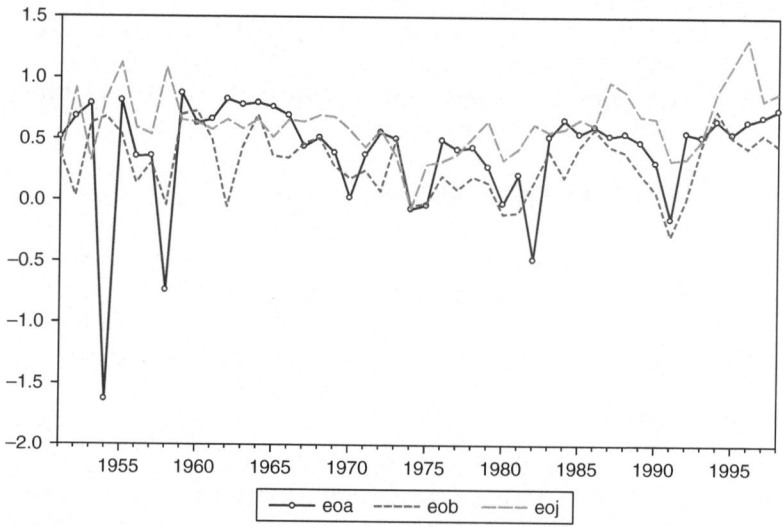

Figure 1.1 Elasticity of real outputs of the three countries

1.2.1 The Determinants of the Inflation Rate

We will now consider the determinants of growth in the price level gP and those of the output growth gQ. In a manner similar to Calvo (1983), Rotemberg (1996), and Gali and Gertler (1999), we consider a firm's quadratic cost function, $C_t = C(s \geq t)$, of the following form, where t means the current year. The firm minimizes the cost function with respect to the price p in logarithm (that is, $p_t = \ln P_t$) that it charges for its product facing a monopolistically competitive market.[1]

$$C_t = E_t \sum_{\substack{s=t \\ s-t}}^{\infty} \delta^{s-t}[(p_s - p_s^*)^2 + k(p_s - p_{s-1})^2],$$

where the first squared term is the out-of-equilibrium cost incurred by the firm because its actual price p_s at time s does not equal equilibrium price p_s^*. (See Equation (1.3) below for the determinants of p_s^*.) The second squared term is the adjustment cost from period $s-1$ to s. All the p_ss and p_s^*s are measured in natural logarithms. The term E_t is an expectation operator at year t, where the random variables are future prices with subscripts $t+j$ ($j \geq 1$). A constant, k, measures the relative weight between the price adjustment cost and out-of-equilibrium cost; a larger k implies a larger adjustment cost relative to the other in the firm's cost calculation. Another constant, δ, is a discount factor taking a value between zero and unity.

Since p_t^* does not depend on p_t (Equation (1.3)), the firm's optimization can be achieved when

$$\frac{\partial C_t}{\partial p_t} = 0 = 2E_t[p_t - p_t^* + k(p_t - p_{t-1}) - \delta k(p_{t+1} - p_t)].$$

It is easily seen that the second-order condition is met. Assuming that the discount factor δ is close to unity, and writing $gP_t = p_t - p_{t-1}$ so that $E_t gP_{t+1} = E_t(p_{t+1} - p_t)$, where, as before, '$g$' stands for 'the growth rate of', the above relationship can be approximated by

$$p_t - p_t^* + k(gP_t - E_t gP_{t+1}) = 0. \tag{1.2}$$

We assume now that the ratio of the firm's optimal price at period t to the competitor's price p_t^c is equal to the log of firm's marginal cost relative to the trend level of the marginal cost, mc_t.[2] Then, recalling that all the ps are the logs of corresponding Ps, one can write this relationship as

$$p_t^* = mc_t + p_t^c + e_t, \tag{1.3}$$

where e_t is an iid (independent and identically distributed) error term with mean zero and a constant variance (that is, white noise; this restriction also applies to other error terms to appear in the following), and

$$mc_t = x_t + h(Q_t), \tag{1.4}$$

where x_t is a log of the nominal wage level relative to its trend, and $h(Q_t)$, a function of output Q_t, is a log of a reciprocal of marginal productivity of labor relative to its trend level. (The assumption that the firm is facing a monopolistically competitive product market might require that $mc_t = p_t^* - p_t^c - e_t$ be augmented by $\ln(1 - \gamma)$ where γ is a reciprocal of the price elasticity of product demand; however, as long as γ remains a constant, ignoring this factor does not cause any problem.) Also, we assume that $h(Q_t)$ will increase in the short-run when some components of aggregate demand increase or technical progress occurs because, in the face of these, output increases temporarily, which will result in increases in the reciprocal of the marginal productivity of labor (on assuming the decreasing marginal productivity of labor in the short-run; that is, $h'(Q_t) > 0$).

Next, substituting (1.3) and (1.4) into (1.2) and solving the resulting relationship for gP_t, one obtains

$$gP_t = E_t gP_{t+1} + \frac{1}{k}(p_t^* - p_t) = E_t gP_{t+1} + \frac{1}{k}[x_t + h(Q_t) + p_t^c + e_t - p_t].$$

When the above relationship is applied to the macro-economy, it would be natural to suppose that the firm's actual price is approximately equal to the competitors' prices, that is, $p_t = p_t^c$. Then the above equation becomes

$$gP_t = E_t gP_{t+1} + \frac{1}{k}[x_t + h(Q_t) + e_t],$$

where k is a positive constant. Changing the notation slightly such that $x_t/k = w_t$, $h(Q_t)/k = l_t$, and $e_t/k = u_t$, the above relationship becomes

$$gP_t = E_t gP_{t+1} + w_t + l_t + u_t, \tag{1.5}$$

where l_t is a function of some aggregate demand components and/or of the 'state of technology', the higher value of which will raise output at least in the short-run.

Finally, we assume that the log of an actual wage level relative to its trend, x_t, is a positive function of some measure of excess demand for labor (see note 2 again).[3] We represent this measure by the rate of change of the unemployment of labor, gur_t (for the USA and the UK)[4] or the ratio of the numbers of job offers to job applications at public employment security offices, roa_t (for Japan; see below for more detail). Hence,

$$w_t = a_0 + a_1' gur_t \text{ (or } a_1' roa_t) + v_t, \tag{1.6}$$

where a_1' is a positive constant and v_t is an iid error term with a zero mean.[5]

Substituting (1.6) into (1.5), and attaching a positive coefficient on $E_t g P_{t+1}$ and l_t, one finds

$$gP_t = a_0 + a_1 E_t g P_{t+1} + a_2 gur_s \text{ (or } a_2 roa_s) + a_3 l_s + \eta_t, \tag{1.7}$$

where the subscript s implies either t (the current year) or $t - j$ ($j > 1$: some past year); the following regression will show which subscript best describes the actual economic process. The term $\eta_t \equiv u_t + v_t$ is also a zero mean iid error term.

1.2.2 The Determinants of Output Growth

We are now concerned with the determinants of output growth in year t as a function of past and contemporaneous macroeconomic variables. Since the explanatory variables are mainly short-run determinants, they consist of demand factors as well as supply factors.

As demand factors, we consider (a) the central government's budget deficits (in ratio form, expenditure/revenue, written as er), (b) current account surpluses (in ratio form, export/import, written as em), and (c) money supply ($m2$) that includes cash in circulation as well as demand and time deposits. Also, the supply factors in our growth equation are (d) the total factor productivity (tf) and (e) the unemployment rate for labor for the USA and UK, or the ratio of job offers to job applications for Japan. The first two factors are written in ratios, which can get round procedures to have constant-price values. The *levels* of those five factors (a) through (e) would affect the output *level* by way of its demand and supply sides; hence it would be their *growth* rates that affect the *growth* of output (*gQ*). If the annual growth rate of *er* is written as *ger*, and so on, then the output growth would be written as

$$gQ_t = b_1 ger_s + b_2 gem_s + b_3 gm2_s + b_4 gtf_s + b_5 gur_s \text{ (or } b_5 oaj_s) + \mu_t,$$
(1.8)

where μ_t is an iid error term with a zero mean, and subscript s in (1.8) implies either t (the current year) or $t-j$ ($j > 1$).[6] The reason for *oaj* entering as a level, not as a growth rate, is that both job offers and job applications are marginal quantities (they do not include demand for and supply of labor who are already employed), so that *roa* itself is close in concept to a growth rate. (See the second paragraph following Japan's estimation of *eo* (1.13).)

Substituting the determinants of *gP* and *gQ*, (1.7) and (1.8), into the definition of *eo* (1.1), while recalling note 6, and then linearizing the resulting expression, one arrives at an equation for *eo* as a function of explanatory variables, which is to be estimated in the next section:

$$eo_t = c_0 + c_1 ger_s + c_2 gem_s + c_3 gm2_s + c_4 gur_s \text{ (or } c_4 roa_s)$$
$$+ c_5 gtf_s + c_6 gPe_{t+1} + \varepsilon_t$$
(1.9)

where $gPe_{t+1} \equiv E_t gP_{t+1}$, which is an expected inflation rate for year $t+1$ formed at year t. The terms c_k ($k = 0$ through 6) are parameters to be estimated. Also, ε_t is a zero mean iid error term.

1.3   Estimation of *eo* (Nominal Aggregate Income Elasticity of Output)

Let us begin with the expectation formation on inflation, gPe_{t+1}. The rational expectations hypothesis implies

$$gP_{t+1} = gPe_{t+1} + v_{t+1},$$
(1.10)

where gP_{t+1} is the actual inflation rate, v_{t+1} is an iid error term with mean zero, and the two terms on the right-hand side are not correlated (that is, $\text{cov}(gPe_{t+1}, v_{t+1}) = 0$); see, for example, Lovell (1986). Substituting gPe_{t+1} from (1.10) into (1.9) yields

$$eo_t = f(s \leq t) + c_6(gP_{t+1} - v_{t+1}) + \varepsilon_t,$$
(1.9')

where $f(s \leq t)$ is the first six terms on the right-hand side of (1.9). If the ordinary least squares method is applied to the above, the resulting estimates are not consistent because gP_{t+1} and v_{t+1} are correlated in view of (1.10). Hence one needs to estimate gP_{t+1} in (1.9') by the instrumental variable method. Instruments were then chosen to be gP_t, gP_{t-1}, gP_{t-2}, for the USA and the UK, while for Japan those variables as well as gP_{t-3} were used as instruments. These instruments yielded a higher adjusted coefficient of determination and better LM statistics (for residual autocorrelation) than other cases where the number of lagged gPs is reduced or extra lagged gPs are added.[7,8] The estimated expected inflation rate in $t + 1$, gPe_{t+1}, was then used in (1.9') in place of gP_{t+1}, and next (1.9) was estimated.[9,10]

To obtain the growth of total factor productivity, *gtf*, consider the following Cobb-Douglas production function of the economy:

$$Q = (tf)(K \cdot U)^{1-\alpha}(N \cdot H)^{\alpha},$$

where *tf* is the level of total factor productivity, K is the aggregate existing stock of capital, U is the utilization rate (operating rate) of capital, N is the number of labor employed, H is hours worked per worker, all in year t (t is omitted for brevity of notation), $1 - \alpha$ is the capital's share in GDP (the elasticity of output regarding capital in use), and α is the labor's share (the elasticity of output regarding labor); hence the constant returns to scale relative to the two factors are assumed. Taking the logarithm of the production function and differentiating it with respect to time, one has

$$gtf = gQ - (1 - \alpha)(gK + gU) - \alpha(gN + gH),$$

where *gtf* is the growth rate of *tf*, and similarly for other terms beginning with g. The data appendix to this chapter describes the data source and more detailed derivation of *gtf*.

The whole period of estimation is from 1951 through 1998. Japan saw consecutive deflationary years in the GDP deflator since 1999. Hence *eoj* also showed erratic behavior in that time (for example, it was 2.845 in 2000, and −136 in 2003). Therefore, we set 1998 as the last year for all three countries.[11] The data are aggregated annually. Because quarterly data on government expenditure and revenue were so far not available, estimation with quarterly data has not been made.

An estimation result for the USA is shown below, where the variables in (1.9) are suffixed by 'a' to remind us that they are 'American' variables. Also, to simplify notation, $gPe_{t+1}a$ is written as $gpea(1)$, which is similarly done for the other countries. In the US case, the White heteroskedasticity test (with cross terms) gives F-statistics $= 3.049$ ($p = 0.017$ with the null-hypothesis of there being homoskedastic error terms). Hence for the USA the White heteroskedasticity consistent covariance matrix was used.[12] (Since χ^2-tests for heteroskedasticity using $n\,\bar{r}^2$, where n is the sample number ($= 48$) and $\bar{r}^2$ is the adjusted coefficient of determination, yield the same results, these test results are omitted here and in footnote 12). The result turned out to be

eoa	const.	gera	gm2a	gema	gura	gtfa	gpea(1)
aver.	0.563	1.298	2.231	0.590	−1.818	−1.944	−6.378
= 0.410	(5.275)	(2.632)	(2.055)	(1.193)	(12.04)	(0.645)	(2.943)

$$(1.11)$$

$$SSR = 1.761, \quad SER = 0.207, \quad \bar{r}^2 = 0.786, \quad DW = 1.779,$$

where the first line below the variables shows the coefficient values, and figures in parentheses are the t-ratios in absolute value. The percentile of the t-distribution with degree of freedom 41 (48–7) at the 95% critical point is 1.683, and that at the 99% critical point is 2.421. Hence all the coefficients except for $gema$ and $gtfa$ are significant at least at the 5 percent significance level in one-sided tests. The 'aver.' below eoa is its sample mean. The lag length of zero or $j (= 1$ or 2) was tried for each of the first five explanatory variables, and the one yielding the highest t-ratio was chosen. SSR is the sum of squared residual, SER is the standard error of regression, $\bar{r}^2$ is the adjusted coefficient of determination, and DW is the Durbin-Watson ratio. This DW-ratio falls in the region where one cannot decide if the errors have autocorrelation, but the LM test (χ^2-test) for autocorrelation with 10 lags yields the p-value of 0.931 (under the null that the errors do not have autocorrelation), so that one cannot reject the null hypothesis.

It is a bit surprising to see that in the US case (only), technical progress, $gtfa$, did not significantly affect eoa nor, therefore, epa ($= 1 - eoa$); this feature remains even when one enters a dummy taking 0 for years 1954 and 1958 as is done in Chapter 3; see Figure 1.1. We will return to this point when the US's *eo* is again estimated under a slightly different assumption. Also, the growth of the export/import ratio, *gema*, is not significant on *eoa* and therefore on *eop*, although the above positive sign is what one expects it to be. For the USA, instead of growth of money supply *m2a*, *gm2a*, growth of money supply *m1a*, *gm1a*, was tried. However, the coefficient of *gm1a* or its lagged value was not significant even at the 10% level.

An estimation result for the UK is shown below, where the variables in (1.9) have a letter '*b*' in their tail to show that they are 'British' variables.

eob	const.	*gerb*	*gm2b*(−1)	*gemb*	*gurb*	*gtfb*	*gpeb*(1)
aver.	0.432	−0.819	−0.677	−0.539	−0.630	4.563	−1.293
= 0.309	(9.939)	(−2.251)	(−2.047)	(−1.280)	(−6.296)	(4.991)	(−2.125)

$$(1.12)$$

$$SSR = 0.783, \ SER = 0.138, \ \bar{r}^2 = 0.710, \ DW = 1.691.$$

Here, as in the USA, the *DW*-ratio does not enable us to decide that the errors do not have autocorrelation. If one turns to the LM test, however, the test with 10 lags yields the *p*-value of 0.141, so one cannot reject the null that the errors do not have autocorrelation. In the UK case, the rate of change in the export/import ratio, *gemb*, is not significant. Also, the increases in rates of change of demand factors (government expenditure/revenue, money supply *m2*) acted to decrease *eo*, that is, they increased the UK's *ep*, *epb* (aggregate income elasticity of the price level; recall *eob* + *epb* = 1). The higher growth of the UK unemployment rate brings down its elasticity *eob*. The faster technical progress in the form of larger *gtfb* made for larger *eob* (that is, smaller *epb*) as one can expect. The higher inflation expectations resulted in smaller *eob*, obviously through larger current inflation rate, *gpb*. For the UK also, the growth of *m1b*, *gm1b* (or its prior value), was not significant on *eob* nor *epb* at the 10% level.

Finally, an estimation for Japan turned out as follows, where the last letters '*j*' stand for 'Japan':

eoj	const.	*gerj*(−1)	*gm2j*(−1)	*gemj*	*roaj*(−1)	*gtfj*	*gpej*(1)	*du*
aver.	1.188	1.394	−1.396	−0.515	−2.723	2.698	−3.211	−0.285
= 0.627	(17.862)	(2.388)	(3.330)	(2.793)	(4.384)	(3.222)	(2.742)	(5.724)

$$(1.13)$$

$du = 1$ for 1974 through 1983, and 1991 through 1993; $du = 0$ in other years. $SSR = 0.505$, $SER = 0.115$, $\bar{r}^2 = 0.793$, $DW = 1.688$.

In Japan, since the inflation immediately after the war was quite high, the expected inflation was severely affected upward. Hence, for Japan only, the estimation starts with 1952.

For this country, one cannot be confident of no serial correlations in error terms from the DW-ratio, but the p-value from the LM test with 10 lags gives a value of 0.897, which allows us to accept the null of no serial correlations in the error terms. In the Japanese case, the regression on the first six variables (without du) has a DW-ratio of 0.991, which obviously indicates the errors display serial correlation. Hence we use a dummy variable du, which is specified as above. The years when $du = 1$ coincide with the two oil shocks, which broke out at the ends of 1973 and 1979, as well as with the early phase of bubble bursting.

For Japan, the rate of change of the unemployment rate, $gurj$, is not used, and instead we use the ratio of the numbers of job offers to job applications at public employment security offices (job placement offices) in Japan, $roaj$. Traditionally, because of Japan's labor market practice of 'long-term employment system', it has been maintained that $roaj$, which is an 'ex $ante$' variable, is a better measure for the labor market tightness. But as the estimation of Chapter 3, which uses $gurj$, will show, $gurj$ yields as significant a coefficient as $roaj$ does. The only difference between the two estimations is that in this chapter $roaj$ has a negative coefficient, but in Chapter 3 $gurj$ takes on a positive coefficient. In other words, $roaj$ represents a constraining supply factor, while $gurj$ expresses an expantionary supply factor. This contrast might be due to the fact that $roaj$ is the ratio between marginal quantities ($roaj$ does not include labor force already employed), while $gurj$ is the growth of the ratio between total quantities (the unemployed/the total labor force).

Also, the rate of change of $roaj$ or its lagged series was not significant at the 10% level, probably because, as already noted, job offers and job applications are both marginal concepts. In Japan's estimation also, the growth of $m1$, $gm1j$ (or its lagged series), did not have a significant coefficient even at the 10% level. The similar results for the three countries suggest that, in the short-run, it is money supply $m2$ rather than $m1$ that has a close relationship with changes in real activity and the price level.

We next turn to examine if the countries in question went through structural breaks (that is if they maintain parameter constancy for the estimation period) using the Cusum of Squares test (see, for example, EViews 6, 2007). The Cusum (Cumulative sum of recursive residuals) of Squares and its 5% significance band is shown in Figure 1.2 for the three countries. The period when the Cusum of Squares goes out of the band indicates the structural break.

According to the figures, the USA had a break during 1969 through 1971. In 1971 the US President Nixon declared New Economic Policy, which led the world to the termination of the Bretton Woods System. Although the UK did not have any break during the sampling period, the Cusum of Squares approached the lower significance line in 1990, which was in the middle of its Big Bang. Finally, Japan's figures indicate that it went through two breaks, 1970 through 1974 and 1986. Note that in Japan's structural break test, the regression equation used does not contain the dummy variable. Japan's breaking periods include Nixon's New Economic Policy (1971), which was called the 'Nixon Shock' in Japan, and the first oil shock of 1973, as well as the initial year of bubble formation (1986).

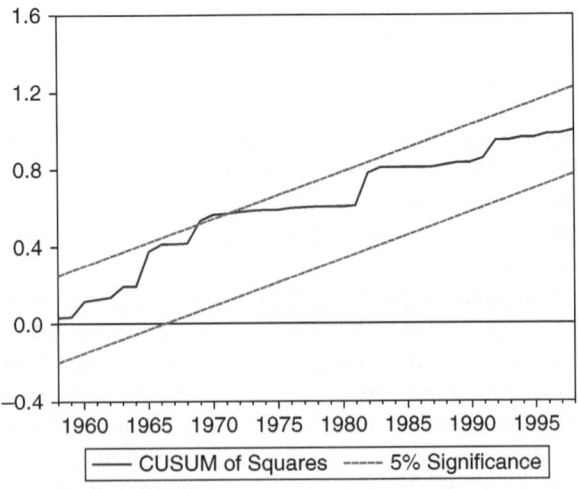

Figure 1.2a The USA

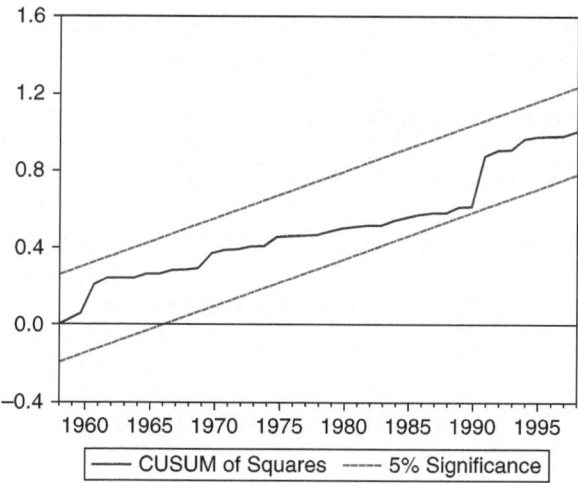

Figure 1.2b The UK

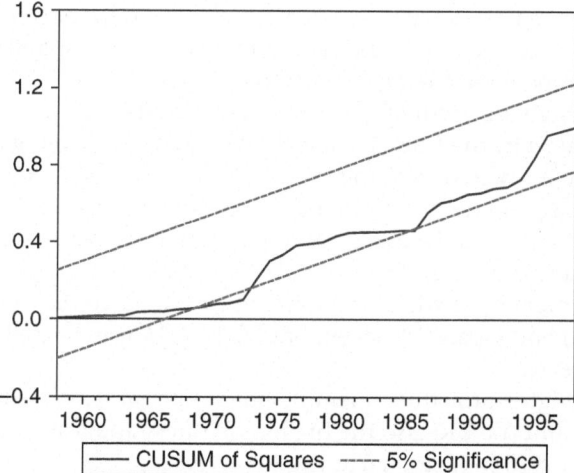

Figure 1.2c Japan

Comparison of the three countries for the whole estimation period gives us some interesting observations as follows:

(i) The sample averages $\overline{eoi}\,(i = a, b, j)$ can be ordered as

$$\overline{eoj}\,(= 0.627) > \overline{eoa}\,(= 0.410) > \overline{eob}\,(= 0.309),$$

where the upper lines imply sample means and, using $\overline{epi} = 1 - \overline{eoi}$,

$$\overline{epb}\,(= 0.691) > \overline{epa}\,(= 0.590) > \overline{epj}\,(= 0.373).$$

Also, one has

$$\overline{eoj}\big/\overline{eoa} = 1.529,\ \overline{eoa}\big/\overline{eob} = 1.327,\ \overline{eoj}\big/\overline{eob} = 2.029.$$

As for the ratios between $\overline{epi}$s:

$$\overline{epb}\big/\overline{epa} = 1.171,\ \overline{epa}\big/\overline{epj} = 1.582,\ \overline{epb}\big/\overline{epj} = 1.853.$$

To given nominal income changes, Japan responded mainly by output changes, the UK mainly by price changes, and the USA fell in between the two countries.

(ii) The higher growth of government expenditure/revenue ratio *ger* raised *eo* in the USA (1% level) and Japan (6% level, $p = 0.051$), while it lowered *eo* in the UK (5% level).

(iii) The higher growth of export/import ratio *gem* reduced *eo* (hence increased *ep*) in Japan (1% level). In the other two countries, the effect was not significant.

(iv) The higher growth of *m2* lowered *eo* (raised *ep*) in the UK (5% level) and Japan (1% level), but in the USA it increased *eo* at the 5% level.

(v) The higher growth of unemployment rate *gur* decreased *eo* in the USA and the UK; also the higher job offers/applications ratio *oaj* lowered *eo* in Japan (all at the 1% level).

(vi) The higher growth of total factor productivity raised *eo* in the UK and Japan (1% level), but the USA saw no significant effect.

(vii) The higher inflation expectation one year ahead had depressing effects on all the *eo*s; that is, in the USA (1% level), the UK (5% level), and Japan (1% level).

1.4   Conclusions

In this chapter we estimated *eo* (nominal income elasticity of output) of three countries, the USA, the UK, and Japan; and implicitly *ep* (the same elasticity of the price level) because of the identity $ep = 1 - eo$.

The estimations for the period 1951 through 1998 gave us interesting contrasts and similarities between the three countries. First, although *eo*s of the three countries responded to variations in the growth of short-run aggregate demand components (including growth of money supply *m2*), in the UK and Japan, those variations were absorbed more in inflation variations, while in the USA, they were absorbed more in fluctuations in output change (so that in the UK and Japan, higher aggregate demands resulted in smaller *eo*s – larger *ep*s –, but in the USA they resulted mainly in larger *eo*).

Secondly, while technical progress contributed to larger *eo*s in the UK and Japan, the effect did not appear in a significant way in the US regression. Finally, inflation expectations one period ahead (forward-looking inflation expectations) had significant negative impacts on *eo*s in all three countries.

Those differences in responses of *eo* and *ep* to shocks of more or less exogenous characters that this chapter considered so far might be related to some structural differences in the economies of the three countries. If this is the case, further inquiry into those differences seems to be an interesting and important agenda for future research.

List of Symbols

eo   aggregate income elasticity of output. (Letters attached to various variables, *a*, *b*, and *j*, mean 'America', 'Britain', and 'Japan', respectively; for example, *eoa*: the US's *eo*).

ep   the same elasticity of the price level.

Y   nominal GDP.

Q   real (constant-price) GDP.

P   price level; $P = Y/Q$.

g   growth operator; for example, *gz*: annual growth rate of *z*.

p   ln*P*, where ln is a natural logarithm.

E   expectation operator.

k	relative weight of price adjustment cost to out-of-equilibrium cost in the firm's objective function.
δ	firm's discount factor.
x	ln(the nominal wage level relative to its trend).
w	x/k.
l	set (*ger gem gm2 gtf*) or its subset.
gur	rate of change of the unemployment rate.
roaj	ratio of job offers to job applications at public employment security offices (job placement offices) in Japan.
ger	growth rate of the central government's expenditure/revenue.
gem	growth rate of (exports of goods and services)/(imports of goods and services).
gm2	growth rate of money supply *m2*.
$gPe_{t+1} = E_t gP_{t+1}$	inflation expectation for year $t+1$ at year t.
K	capital stock at the start of year t.
U	capital utilization rate.
N	number of workers employed.
H	hours worked weekly (manufacturing workers for the USA; male workers of all industries for the UK; hours worked monthly, the average of all workers in all industries for Japan). (Since what matters is the growth rate of H, the length of the sample period, be it a week or a month, is not relevant.)
α	labor's share in national income.
$1-\alpha$	capital's share in national income.

Data Appendix

National income statistics were drawn from *Survey of Current Business* for the USA; *Economic Trends* and *The United Kingdom National Accounts* for the UK; and *National Income Statistics Annuals* of Economic Planning Agency for Japan.

Labor statistics were drawn from *Monthly Labor Review* for the USA; *Monthly Digest of Statistics* for the UK; and *Survey Report on Labor Force* for Japan.

Monetary statistics came from *Federal Reserve Bulletins* for the USA; *Financial Statistics* for the UK. Money supply $m1$ of the UK for the period 1960–1995 was derived from Gordon (2003, Appendix B); and *Bank of Japan Economic Statistics Annuals* for Japan.

For the USA and UK, α was computed from α = (employees' compensation)/(employees' compensation + incorporated business income) using national income statistics data. For Japan, also using national income statistics data, α = [employees' compensation $+Z\cdot$ (personal business income)]/(nominal GNP), where Z = (private and public incorporated business employees' compensation)/[incorporated business employees' compensation + (incorporated business income)]; that is, the unobservable labor's share in personal business income was assumed to be the same as that in private and public incorporated business income. Slightly different formulas were used for α for the USA/UK and Japan, because the classifications of recipients of national income were not continuous over time in the first two countries.

The capital stock figures for the USA and the UK are those computed by O'Mahony (1996) through 1989, and these are linked with the figures in *Survey of Current Business* and *Economic Trends*, respectively. In deriving the growth of total factor productivity, the capital utilization rates of the USA and the UK were assumed to be unity because the figures were not available in the latter two countries. (The UK has not published the figures except for the proportion of plants in full operation in the manufacturing industry.)

The capital stock estimates for Japan were drawn from *National Income Statistics Annuals* of Economic Planning Agency.

Notes

1. The firm's optimal price p_t^* is determined from the equality of marginal cost and marginal revenue, (1.3); and the minimization of C_t decides the speed with which the firm aims for p_t^*.
2. See Gali and Gertler (1999, p.199 et seq.) for similar assumptions: they consider percentage deviations of firms' marginal cost from the steady state value and of aggregate output from its natural level, both as determinants of price inflation.
3. As w_t is the gap between the actual wage level and its trend level, it is a variable conceptually similar to the rate of change of the wage level.
4. As our estimation shows below, the rate of change of the unemployment rate explains the behavior of w_t better than the level of the unemployment

does. One reason for this would be that the wage level is taken as the ratio to its trend level.

5. One might consider x_t to depend on $E_t g P_{t+1}$, whose coefficient would be unity according to the natural rate hypothesis. However, on a short-run horizon, the coefficient of this expectation term may well be smaller than unity. In any case, the expectation term also appears as a first term in (1.5). Hence, neglecting the term in (1.6) does not cause any problem in the following discussion.

6. An explanatory variable l_s in (1.7) is actually a vector, or a subset, of (ger_s gem_s $gm2_s$ gtf_s).

7. However, equalizing the lag number to either three or four for the three countries yielded similar results.

8. Estimation using lagged values of $gm2$, as well as gP, as instruments, was also tried. Those extra instruments, however, yielded lower adjusted coefficients of determination and/or worse LM statistics.

9. This means that we use the two-stage least squares method.

10. Alternatively, using another definition of rational expectations, $gPe_{t+1} = E_t(gP_{t+1}|I_t)$, where I_t is the information available at t, one may regard the estimated value of the above autoregression as gPe_{t+1} and proceed in the same way as in the text. See Maddala (2001, pp. 419–22).

11. If one wants to extend the period of analysis to more recent years, quarterly frequency data would be necessary. In that case, however, one faces a problem of getting quarterly figures of government revenues and expenditures. See the text below.

12. As is shown below, the UK and Japan's White tests (with cross terms) show no heteroskedasticity; hence the White matrix for these two countries was not used. For the UK, $F = 0.456$ ($p = 0.971$) and for Japan, $F = 0.971$ ($p = 0.553$).

References

Calvo, G.A. (1983) Staggered prices in a utility maximizing mramework. *Journal of Monetary Economics* 12, 383–98.

EViews 6 (2007) *User's Guide*. Irvine, CA: Quantitative Micro Software.

Friedman, M. (1970) A theoretical framework for monetary analysis. *Journal of Political Economy* 78, 193–238.

——(1971) A monetary theory of nominal income. *Journal of Political Economy* 79, 323–37.

Gali, J. and M. Gertler (1999) Inflation dynamics: a structural econometric analysis. *Journal of Monetary Economics* 44, 195–122.

Gordon, R.J., ed. (1974) *Milton Friedman's Monetary Framework*. Chicago: University of Chicago Press.

——(2009) *Macroeconomics*, 11th ed. New York: Pearson Education.

Keynes, J.M. (1936) *The General Theory of Employment, Interest, and Money*. London: Macmillan.

Laidler, D. (1973) The influence of money on real income and inflation: a simple model with some empirical tests for the United States, 1953–72. *Manchester School* 41, 367–95.

Lovell, M.C. (1986) Tests of the rational expectations hypothesis. *American Economic Review* 76, 110–124.

McCallum, B.T. (1973) Friedman's missing equation: another approach. *Manchester School* 41, 311–28.

Maddala, G.S. (2001) *Introduction to Econometrics*, 3rd ed. Chichester and New York: Wiley.

Nobay, A.R. and H.G. Johnson (1977) Monetarism: A historic-theoretic perspective. *Journal of Economic Literature* 15, 470–85.

O'Mahony, M. (1996) Measures of fixed capital stocks in the post-war period: a five-country study. In B. van Ark et al., eds. *Quantitative Aspects of Post-War European Economic Growth*, pp. 165–214. Cambridge: Cambridge University Press.

Rotemberg, J.J. (1996) Prices, output, and hours: an empirical analysis based on a sticky price model. *Journal of Monetary Economics* 37, 505–33.

2
Estimating the 'Missing Equations' for Developing Countries

2.1 Introduction

This chapter will discuss how annual nominal income change is divided between output change and the price (level) change. The concept is phrased alternatively as the nominal income elasticities of output (eo) and the price level (ep), respectively, where $eo + ep = 1$, as is shown later on. Also, the elasticities of positive fraction correspond to the intermediate situations between the two cases with which Friedman (1970) closed his macro models: quantity theory and income-expenditure theory. Quantity theory and income-expenditure theory are represented by $eo = 0$ and $eo = 1$, respectively. Here, one cannot fail to mention the concept of eo and ep in Keynes (1936, chs 20 and 21).

Later than Keynes (1936) and Friedman (1970), Friedman (in Gordon, ed. 1974, p. 45), Nobay and Johnson (1977), Laidler (1995) and Gordon (2009) drew attention to the division of nominal income change into changes in output and the price level. It will now be worth referring to a sentence which is part of a personal letter from Friedman to Laidler, and which is mentioned in Laidler (1995, p. 338, note 16):

'... there is still no satisfactory solution to ...how to predict the fraction of a change in nominal income that will take the form of a change in prices rather than in output.'

In this chapter we first formulate firms' optimal pricing policy and then add some simple theory explaining growth of the economy to arrive at, in a reasonably general setting, an equation describing explanatory variables of eo and ep.

Then, using the formula for eo, we estimate the coefficients for nine countries. These are: Chile, India, Korea, Malaysia, South Africa, Spain, Thailand, Turkey, and Venezuela. It should be noted that Korea and Spain are fully-fledged developed countries and OECD (Organisation for Economic Co-operation and Development) members; however, for the sake of comparison and because developing countries in their initial stages often have data problems, we estimate these two countries together with the others. The ranges of estimation are from the latter half of the 20th century to the early 21st century, but exact periods differ among the countries depending on the data availability.

The next section 2 derives the equation for eo based on firms' behavior. Section 3 estimates eos for the above countries, and compares the results among those countries. Section 4 concludes the chapter.

2.2 Firm behavior and division of nominal income between output and prices

Firms are assumed to make a two-stage optimization regarding output and the price level. Let us start with the descriptions of variables and concepts: Y represents nominal income, Q real GDP, and P the price level. Then we have $Y = P \cdot Q$. Letting gz be an annual growth rate of z, we at once find

$$gY = gP + gQ.$$

We now define eo as the elasticity of output regarding nominal income, and ep as the elasticity of the price level regarding nominal income. Then we have

$$eo = \frac{\Delta Q}{\Delta Y} \frac{Y}{Q} = \frac{\Delta Q/Q}{\Delta Y/Y}$$

where Δ is a difference operator. If we regard $\Delta Y_t = Y_t - Y_{t-1}$, where subscript t refers to some year, then we have

$$eo = \frac{gQ}{gP + gQ} = \frac{1}{1 + \frac{gP}{gQ}}, \quad ep = \frac{gP}{gP + gQ}. \tag{2.1}$$

Hence one immediately finds $eo + ep = 1$.

The rest of this chapter, therefore, will pay attention to the determinants of gQ and gP to look for the variables on which eo will depend.

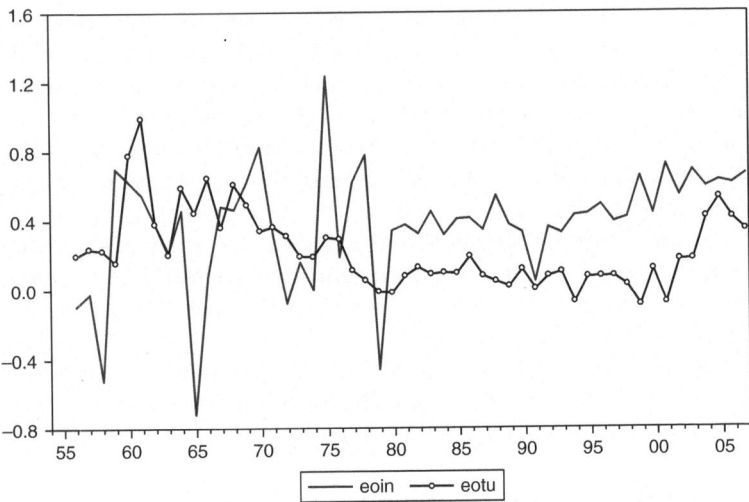

Figure 2.1 Elasticity of real outputs of India and Turkey

We will deal with nine (mainly) developing countries, using postwar annual data; these are Chile, India, Korea, Malaysia, South Africa, Spain, Thailand, Turkey, and Venezuela. Annual changes in *eo* of India and Turkey are shown in Figure 2.1 (some countries include *eos* with abnormally high or low values, so these *eos* are not suitable to be shown; *eo* and *ep* are supposed to lie in the range $0 \le eo, ep \le 1$).

2.2.1 The determinants of inflation rates

We start with the determinants of the growth of the price level *gP* and of total output *gQ*. In a similar way to Calvo (1983), Rotemberg (1996), and Gali and Gertler (1999) we consider a firm's cost of adjustment $C_t = C(s \ge t)$, the cost being entailed in price changes, of the following form, where *t* means the current year. This cost function is the same as in Chapter 1, and hence the description will be made simpler. The firm minimizes the cost function regarding price p_t in logarithm (that is, $p_t = \ln P_t$) that it charges for its product facing a monopolistically competitive market.[1]

$$C_t = E_t \sum_{s=t}^{\infty} \delta^{s-t} [(p_s - p_s^*)^2 + k(p_s - p_{s-1})^2],$$

where the first squared term is the out-of-equilibrium cost incurred by the firm because its actual price p_s at time s does not equal equilibrium price p_s^*. (See equation (2.3) below for the determinants of p_s^*.) The second squared term is the adjustment cost from period $s-1$ to s. All the p_ss and p_s^*s are measured in natural logarithms. The term E_t is an expectation operator at year t, where the random variables are future prices with subscripts $t+j$ ($j \geq 1$). The firm's cost minimization can be achieved when

$$\frac{\partial C_t}{\partial p_t} = 0 = 2E_t[p_t - p_t^* + k(p_t - p_{t-1}) - \delta k(p_{t+1} - p_t)].$$

Assuming that the discount factor δ is close to unity, and writing $gP_t = p_t - p_{t-1}$, so that $E_t gP_{t+1} = E_t(p_{t+1} - p_t)$, where, as before, '$g$' stands for 'the growth rate of', the above relationship can be approximated by

$$p_t - p_t^* + k(gP_t - E_t gP_{t+1}) = 0. \tag{2.2}$$

We now assume that the ratio of the firm's optimal price at period t to the competitor's price p_t^c is equal to the log of the firm's marginal (production) cost to its trend level, mc_t.[2] Then, recalling that all the ps are the logs of the corresponding Ps, one can write this relationship as

$$p_t^* = mc_t + p_t^c + e_t \tag{2.3}$$

where e_t is an iid (independent and identically distributed) error term with mean zero and a constant variance (that is white noise; we also apply this assumption to other error terms to appear in the following), and

$$mc_t = x_t + h(Q_t) \tag{2.4}$$

where x_t is a log of the nominal wage level relative to its trend, and $h(Q_t)$, a function of output Q_t, is a log of a reciprocal of marginal productivity of labor relative to its trend level. (The assumption that the firm is facing a monopolistically competitive product market might require that $mc_t = p_t^* - p_t^c - e_t$ be increased by $\ln(1 - \gamma)$ where γ is a reciprocal of the price elasticity of product demand; however, as long as γ remains a constant, ignoring this factor will not cause any problem.) Also, we assume that $h(Q_t)$ will increase in the short-run when some components of aggregate demand increase, because if they happen, they will lead to temporary increases in output, which will result in increases in the reciprocal of marginal productivity of labor (on assuming the decreasing marginal productivity of labor in the short-run; that is, $h'(Q_t) > 0$). In the case of technical progress, $h(Q_t)$ will decrease.

Now, from (2.2), (2.3), and (2.4) we obtain

$$gP_t = E_t gP_{t+1} + \frac{1}{k}(p_t^* - p_t) = E_t gP_{t+1} + \frac{1}{k}[x_t + h(Q_t) + p_t^c + e_t - p_t].$$

When the above relationship is applied to the macro-economy, it would be natural to suppose that the firm's actual price is approximately equal to the competitors' prices, that is, $p_t = p_t^c$. Then the above equation becomes

$$gP_t = E_t gP_{t+1} + \frac{1}{k}[x_t + h(Q) + e_t],$$

where k is a positive constant. Changing the notation slightly so that $x_t/k = w_t$, $h(Q_t)/k = l_t$, and $e_t/k = u_t$, the above relationship becomes

$$gP_t = E_t gP_{t+1} + w_t + l_t + u_t \tag{2.5}$$

where l_t is a function of some aggregate demand components and/or of the 'state of the technology'; the higher value of the former (latter) will raise (lower) the inflation at least in the short-run.

In Chapters 1 and 3, where we deal with OECD countries, we use unemployment rates or their rates of change as determinants of wages. But for most countries that are the subjects of this chapter, series on unemployment rates are generally not available, so that in the following we use 'business cycle factors' to represent the tightness of labor markets as well as the speed of technical progress. Business cycle factors are represented by the residuals of OLS regressions of ln(real GDP) on a constant, a trend, and a trend squared, where a trend is just the number applied to each year in order (for 1955 it is 1). A positive residual implies that in those periods, business activity is more vibrant, labor markets are tighter, and/or technical progress is faster than normal years.

Hence in Equation (2.5), we substitute res_t, representing the residual of the above regression, for w_t. Then attaching a coefficient on $E_t gP_{t+1}$, res_s, and l_s, Equation (2.5) becomes

$$gP_t = a_0 + a_1 E_t gP_{t+1} + a_2 res_s + a_3 l_s + v_t, \tag{2.6}$$

where subscript s implies either t (the current year) or $t-j$ $(j > 1)$: some past year; the following regression will show which subscript best describes the actual economic process. The term v_t is also a zero mean iid error term.

2.2.2   The determinants of output growth

We will next consider the determinants of output growth in year t as a function of past and current macroeconomic variables. Since the explanatory variables are mostly short-run determinants, they mainly consist of demand factors as well as a supply factor.

As demand factors, we consider (a) current account surpluses (in ratio form, export/import, written as em), (b) the central government's budget deficits (in ratio form, expenditure/revenue, written as er), (c) money supply ($m1$, $m2$, or mq; $m1$ consists of cash in circulation and demand deposits, $m2 = m1 +$ time deposits, and mq is made up of time deposits; mq is applied to Turkey only), (d) business cycle factors, which are represented by surpluses of actual GDP over its trend in real GDP. These surpluses might include technical progress (as well as business intensity) which is faster than the average represented by the trend. Interpreted in this way, positive surpluses can be a supply factor giving rise to a surge in GDP.

The levels of these four factors would affect the output level by way of its demand and supply sides; hence it would be their growth rates that affect the growth of output (gQ). If we write the annual growth rate of er as ger, and so on, then the output growth would be written as

$$gQ_t = b_1 gei_s + b_2 ger_s + b_3 gmi_s + b_4 res_s + \mu_t, \qquad (2.7)$$

where μ_t is an iid error term with a zero mean. The first g implies, as before 'the growth rate of'. In the case of res, it explains gQ_t better than $gres$ does, except for Venezuela, probably because res implies *divergence* of Q_t from its trend. The term i in gmi_s is one of 1, 2, or q. A subscript s in (2.7) implies either t (the current year) or $t-j(j > 1)$.[3]

Substituting the determinants of gP and gQ, (2.6) and (2.7), into the definition of eo (2.1), referring to note 3, and then linearizing the resulting expression, we have an equation for eo as a function of explanatory variables, which is to be estimated in the next section:

$$eo_t = c_0 + c_1 gei_s + c_2 gem_s + c_3 gmi_s + c_4 res_s + c_5 gPe_{t+1} + \varepsilon_t \qquad (2.8)$$

where $gPe_{t+1} \equiv E_t gP_{t+1}$, which is an expected inflation rate for year $t+1$ formed at year t. The terms c_k ($k = 0$ through 5) are the parameters to be estimated. Also, ε_t is a zero mean iid error term.

Before moving on to the next section, we shall discuss expectation formation on inflation gPe_{t+1}. The rational expectations hypothesis implies

$$gP_{t+1} = gPe_{t+1} + \lambda_{t+1}, \tag{2.9}$$

where gP_{t+1} is the actual inflation rate, λ_{t+1} is an iid error term with mean zero, and the two terms on the right-hand side are not corre-lated (that is, $\text{cov}(gPe_{t+1}, \lambda_{t+1}) = 0$). See, for example, Lovel (1986). Substituting gPe_{t+1} from (2.9) into (2.8) yields

$$eo_t = f(s \leq t) + c_5(gP_{t+1} - \lambda_{t+1}) + \varepsilon_t \tag{2.9'}$$

where $f(s \leq t)$ represents the first six terms on the right-hand side of (2.8). If the OLS is applied to the above, the resulting estimates are not consistent because gP_{t+1} and λ_{t+1} are correlated in view of (2.9). Hence we need to estimate gP_{t+1} in (2.9) by the instrumental variable method. Actually, however, since people cannot know gP_{t+1} in period t, we regress gP_t on the past two gP_ts, and shift it one period ahead to regard it as gPe_{t+1}. Instruments were then chosen to be gP_{t-1} and gP_{t-2}. These instruments yielded a higher adjusted coefficient of determination and better LM statistics (for residual autocorrelation) than other cases where the number of lagged gPs is reduced or extra gPs are added.[4] The estimated expected inflation rate in $t+1$, gPe_{t+1}, was then used in (2.9') in place of gP_{t+1}, and we next estimated (2.9).[5,6] In the next section, gPe_{t+1} is written as $gpe(1)$.

2.3   Estimation of *eos* for nine countries

We shall deal with the nine countries in the alphabetical order. The annual data were drawn from *International Financial Statistics*, Interna-tional Monetary Fund, 1979, 1993, 2003, and 2008. The first country is Chile. In (2.10) the first row is the dependent and independent variables. The second row is estimated coefficients, and the third row shows p-values (significance levels for the t-distribution). Each esti-mation includes a constant, but it is not shown. In the Chilean and South African cases only, the business cycle factor is not significant even at the 10% level. The figures in the following parentheses are the period of estimation and effective observation number: (1965–2007; 43)

eo	gei(−1)	ger(−1)	gm2(−1)	gpe(1)	res	du
	−0.494	−0.822	−0.103	−0.971	0.378	17.895
	(0.035)	(0.026)	(0.022)	(0.004)	(0.298)	(0.000)

$$(2.10)$$

$\overline{r^2} = 0.995,\ SER = 0.184,\ \overline{eo} = 0.223,\ DW = 1.375,\ LMP = 0.206.$

Here $\overline{r^2}$ is the adjusted coefficient of determination, SER is the regression standard error, figures in parentheses are p-values, $\overline{eo}$ is the sample mean, DW is the Durbin-Watson ratio, and LMP is a p-value for Breush-Godfrey's serial correlation LM test, with the null being that there is no serial correlation. The dummy variable takes on 1 in 1982 (when $eo = 17.691$; normally eo is expected to fall within $0 \leq eo \leq 1$); in other years $du = 0$. When computing $\overline{eo}$, we exclude the year with a dummy variable. In the Chilean case the first three factors, enhancing demand, raise inflation relative to output growth (see the second equation contained in Equation (2.1)). The factor $gm2$ is the growth rate of $m2$ consisting of time deposits as well as cash and demand deposits.

For India, the estimation turned out as (1958–2006; 49)

eo	ge	ger	gm1(−1)	gpe(1)	res	du
	−0.463	−0.345	−0.630	−0.006	1.597	−0.891
	(0.002)	(0.043)	(0.079)	(0.884)	(0.039)	(0.000)

$$(2.11)$$

$\overline{r^2} = 0.868,\ SER = 0.123,\ \overline{eo} = 0.400,\ DW = 1.865,\ LMP = 0.682.$

Dummy du takes on 1 in 1958, 1965, and 1979, when $eo = -0.522$, -0.718, and -0.458, respectively.

In India's case, money supply $m1$ (excluding time deposits) provides better results. The main difference between Chile and India is that in the latter, the business cycle factor raises eo significantly.

The third country we deal with is Korea. It was estimated as (1957–2007; 51)

eo	gei(−1)	ger(−1)	gm2(−1)	gpe(1)	res	du
	−0.178	0.216	−0.176	−3.485	3.501	4.120
	(0.013)	(0.049)	(0.070)	(0.000)	(0.000)	(0.000)

$$(2.12)$$

$\overline{r^2} = 0.971$, $SER = 0.100$, $\overline{eo} = 0.494$, $DW = 1.512$, $LMP = 0.149$.

Dummy *du* takes on 1 only in 1998, when *eo* = 4.278. An interesting feature of this country is that larger growth of government deficits raises *eo* (that is, raises output growth relative to inflation).

We now turn to Malaysia, which is estimated as (1958–2007; 50)

eo	gei(−1)	gm2(−1)	gpe(0)	res	du	
	−0.243	−0.876	−6.593	1.506	−14.888	(2.13)
	(0.567)	(0.096)	(0.083)	(0.041)	(0.000)	

$\overline{r^2} = 0.983$, $SER = 0.282$, $\overline{eo} = 0.637$, $DW = 1.896$, $LMP = 0.926$.

Malaysia's data from *IMF Financial Statistics* have the government expenditure/revenue ratio only for 1961 through 1999, and the growth rate of this ratio (*ger*) did not turn out to be significant, hence we deleted *ger* from the estimation. Only in Malaysia, the estimated inflation rate on the current year, which was regressed over the past two years, appears as a (barely significant) expected inflation variable, taking on a correct sign. For this country, the dummy takes on 1 in 1957 and 1998 (other years being zero), where *eo* in 1957 is −1.394, while in 1998 it is −14.322.

We will next look into South Africa, which yields (1958–2006; 49)

eo	gei	ger	gm2(−1)	gpe(1)	res	du	ar(1)
	−0.420	−0.335	−0.537	−3.653	0.796	0.219	0.567
	(0.000)	(0.153)	(0.015)	(0.000)	(0.121)	(0.053)	(0.000)

$$(2.14)$$

$\overline{r^2} = 0.838$, $SER = 0.111$, $\overline{eo} = 0.299$, $DW = 2.037$, $LMP = 0.155$.

The above estimation assumes that the errors follow a first-order autoregressive process, but without *ar*(1), $DW = 1.270$, and $LMP =$

0.030, which obviously shows that serial correlation occurs in error terms. Dummy *du* takes on 1 in 1963 only, when $eo = 1.080$.

For this country, we find that growth of trade deficits and of money supply *m2* affect inflation positively, and they affect real growth negatively, in a significant way.

Spain's *eo* can be explained mainly by inflation expectation, money supply, and the business cycle factor, as follows:
(1964–1998; 35)

eo	*gei*	*ger*	*gm1*	*gpe*(1)	*res*	*du*	*ar*(1)
	−0.023	−0.559	0.747	−3.515	0.977	−0.488	0.443
	(0.880)	(0.180)	(0.024)	(0.000)	(0.038)	(0.000)	(0.017)

$$(2.15)$$

$\overline{r^2} = 0.840$, $SER = 0.085$, $\overline{eo} = 0.349$, $DW = 2.043$, $LMP = 0.683$.

When $ar(1)$ is absent, $DW = 1.122$, and $LMP = 0.060$, showing that one cannot exclude the possibility of serial correlation. Here money supply is measured by *m1*. The dummy *du* takes on 1 in two years, 1959, and 1993, where *eos* are −0.494 in 1959, and −0.366 in 1993. It is interesting to note that this is the only country where the growth of monetary aggregate enhances the real output growth relative to inflation.

Thailand's estimation yielded:
(1957–2004; 48)

eo	*gei*(−1)	*ger*(−1)	*gm2*(−1)	*gpe*(1)	*res*	*du*
	−0.03	0.296	−0.457	−4.034	3.135	3.469
	(0.075)	(0.382)	(0.002)	(0.019)	(0.006)	(0.030)

$$(2.16)$$

$\overline{r^2} = 0.686$, $SER = 0.495$, $\overline{eo} = 0.673$, $DW = 2.043$, $LMP = 0.779$.

The dummy takes on 1 only in1961, when $eo=1.016$. Money supply and the business cycle factor make for the two elasticities in different directions; money raises inflation but business upswings (relative increases in aggregate demand) raise output growth rather than inflation. In estimating *res* and *eo* we used the White heteroskedasticity consistent covariance matrix because without it, the *p*-value of the White heteroskedasticity test (the null being no heteroskedasticity) = 0.000.

Although Turkey's estimation period is relatively short, it yielded nice results:

(1970–1995; 26)

eo	gei(−1)	ger(−1)	gmq	gpe(1)	res(−1)	ar(1)	
	−0.093	−0.932	−0.039	−0.296	−0.932	0.609	(2.17)
	(0.014)	(0.048)	(0.015)	(0.001)	(0.009)	(0.001)	

$\bar{r}^2 = 0.839$, $SER = 0.048$, $\overline{eo} = 0.135$, $DW = 1.845$, $LMP = 0.385$.

In Turkey's case, the monetary aggregate is measured by quasi-money, so that it does not include cash in circulation and demand deposits. The estimation of it does not need a dummy variable.

All the demand factors, which are the first three explanatory variables as well as inflation expectation and business cycle factors, make for larger inflation relative to larger output growth. (Without $ar(1)$, $DW = 0.977$.)

We finally turn to Venezuela, which yielded

(1958–2006; 49)

eo	gei(−1)	ger(−1)	gm2	gpe(1)	gres	du	ar(1)
	−1.542	−3.789	2.330	−3.594	−0.170	11.641	0.379
	(0.041)	(0.016)	(0.291)	(0.331)	(0.004)	(0.000)	(0.032)

$$(2.18)$$

$\bar{r}^2 = 0.694$, $SER = 1.70$, $\overline{eo} = 0.322$, $DW = 1.905$, $LMP = 0.121$.

Dummy *du* takes on 1 for three years, 1959, 1960, and 1983, whose *eos* are 2.017, 9.277, and 20.400, respectively. Only in this country, the growth rate of *m2* affect real growth relative to inflation positively, and *changes* in business cycle factors affect real growth relative to inflation negatively (that is, faster cyclical upturns [increases in *gres*] enhances inflation relative to output growth).

Table 2.1 exhibits the coefficients of explanatory variables except for *du* and *ar*(1) because these two do not have much economic meaning. The symbols °, *, and ** attached to the coefficients mean they are significant at the 10, 5, and 1% levels, respectively. Those without any signs are insignificant at the 10% level. See notes to the table for further descriptions.

The first point one notes is that, among the significant coefficients, increases in growth of export/import, government expenditure/revenue, and money supply all raise inflation rather than real

Table 2.1 Regression coefficients

	gei	ger	gmi	gpe(1)	res
Chile	−0.495*	−0.822*	−0.103	−0.971**	0.378
India	−0.463**	−0.345*	−0.630°	−12.158**	1.597*
Korea	−0.178*	0.216*	−0.176°	−3.485**	3.501**
Malaysia	−0.243	–	−0.876°	−6.593°	1.506*
South Africa	−0.420**	−0.335	−0.537*	−3.653**	0.796
Spain	−0.023	−0.559	0.747*	−3.515**	0.977*
Thailand	8.07E−5	−0.165	−0.561*	−0.787**	0.640*
Turkey	−0.093*	−0.145*	−0.039*	−0.296**	−0.932**
Venezuela	−1.541*	−3.789*	2.320	−3.594	−0.170**

Notes: (1) The first three variables may have one year lags. See the above individual examinations. (2) The concept of 'money' is not the same across countries. See also the above individual cases. (3) For Malaysia, *ger* is dropped because of the lack of data.

growth rates (except for *ger* of Korea). This is reflected by the fact that the nine country average of *eo*, $\overline{eo}$, excluding the years with dummies, is 0.384, while the corresponding *ep*, $\overline{ep}$, is 0.616. In other words, prices are much more responsive to outside demand factors than outputs are.

The next point to note is that in five countries out of seven with significant results, business cycle factors (deviations of output from its trend) are growth-promoting rather than inflation-inviting; the exceptions are Turkey and Venezuela, and also note that in Venezuela, it is higher growth of the business cycle factor that lowers real output growth (raises inflation).

Finally, and incidentally, the order correlations between $\overline{eo}$ and $\overline{gQ}$, which is written as $orcor(\overline{eo}, \overline{gQ})$, where $\overline{gQ}$ is a sample average of growth rate of each country, and between $\overline{ep}$ and $\overline{gP}$, written *orcor* $(\overline{ep}, \overline{gP})$ are $orcor(\overline{eo}, \overline{gQ}) = 0.633$ and $orcor\ (\overline{ep}, \overline{gP}) = 0.767$.[7]

2.4 Conclusions

Starting with two-stage optimization of firms, we derived nominal income elasticities of output and the price level (*eo* and *ep*; $eo + ep = 1$), and then estimated *eo* for nine (mainly) developing countries. The results are summarized in Table 2.2. It will be convenient to further look into the table.

All of the seven significant coefficients of *gei* are negative, implying that increasing growth of export/import ratios reduce output growth relative to inflation (or it increases inflation relative to output growth; see the first relation in Equation (2.1)). Five (out of six) significant coefficients of *ger* are also negative (the exception is Korea) which means that larger growth of government expenditure/revenue ratios tends to decrease output growth relative to inflation.

Further, seven (out of eight) significant coefficients of money growth are negative (the exception is Spain), so that larger *gmi* made for larger inflation relative to output. In other words, increases in the growth of the above three demand elements are inflationary rather than real growth-promoting.

On the other hand, larger business cycle factors, which mean the larger gaps between real output and its trend, are helpful for output growth compared to inflation, in five (out of seven) significant coefficients (the exceptions are Turkey and Venezuela; note that in Venezuela, the negative sign appears on the growth of *res*).

The clear contrast between the signs on the first three variables and those of the last variable seems quite interesting and would merit further analysis and attempts at economic interpretations. Positive signs of *res* might largely reflect supply factors, such as larger employment and faster technical progress, because the first three variables obviously have demand-side characters.

List of Symbols

Y	nominal income.
Q	real income (output).
P	price level (GDP deflator).
eo	nominal income elasticity of output; $d\ln Q/d\ln Y$.
ep	the same elasticity of the price level; $d\ln P/d\ln Y$.
$\overline{eo}$	sample mean of eo of each country, excluding years with dummy variables. $\overline{ep}$, $\overline{gP}$, and $\overline{gQ}$ are defined in similar manners, but $\overline{gP}$ and $\overline{gQ}$ include all the sampled years.
gei	growth rate of the export/import ratio.
ger	growth rate of government expenditure/revenue.
gmi	growth rate of money supply mi, where $m1$ consists of cash and demand deposits; $m2$ of $m1$+time deposits; and mq of time deposits; mq is used for Turkey only.

gpe(1)   expected inflation rate at year *t* for inflation in *t* + 1.

res   'business cycle factor', which is a residual in a regression of $\ln Q$ on a constant, a trend and a trend squared, where a trend is the number applied to each year in order, with $1955 = 1$, and $2007 = 53$.

du   dummy variable attached to years when *eo* takes on abnormal values; normal values lie largely in the region $0 \leq eo \leq 1$.

LMP   *p*-value (significance level) in the Breush-Godfrey Lagrange-multiplier test, under the null of no serial correlation.

Notes

1. As will become clear shortly, since the firm's optimal price in year *t*, p_t^* does not depend on p_t, p_t is a cost minimizing short-term price, while p_t^* is a profit maximizing medium- or long-term price.
2. See Gali and Gertler (1999, p. 199 et seq.) for similar assumptions: they consider percentage deviations of firms' marginal cost from the steady state value, and of aggregate output from its natural level, both as determinants of price inflation. See also Gali (2008, for example, p. 18).
3. An explanatory variable l_t in (2.5) is actually a vector, or a subset of (ger_t gem_t gmi_t gtf_t).
4. We also tried estimation using, as instruments, lagged values of *gPs* as well as *gmi*. These extra instruments, however, yielded lower adjusted coefficients of determination and/or worse LM statistics.
5. This means that we use the two-stage least squares method.
6. Alternatively, using another definition of rational expectations, $gPe_{t+1} = E_t(gP_{t+1}|I_t)$, where I_t is the information available at *t*, one may regard the estimated value of the above autoregression as $gP_{e,t+1}$ and proceed in the same way as in the text. See Maddala (2001, pp. 419–22).
7. The orders of ($\overline{eo}, \overline{ep}, \overline{gQ}, \overline{gP}$) among the nine countries (Chile, India, Korea, Malaysia, South Africa, Spain, Thailand, Turkey, Venezuela), with 1 taking the highest value are, respectively, (7, 3, 2, 1, 6, 5, 8, 9, 4), (3, 7, 8, 9, 4, 5, 2, 1, 6), (7, 5, 1, 2, 9, 6, 3, 4, 8), and (1, 8, 5, 9, 7, 6, 3, 2, 4).

References

Calvo, G. (1983) Staggered prices in a utility maximizing framework. *Journal of Monetary Economics* 12, 383–98.

Friedman, M. (1970) A theoretical framework for monetary analysis. *Journal of Political Economy* 78, 193–238.

Gali, J. (2008) *Monetary Policy, Inflation, and the Business Cycle.* Princeton: Princeton Univrsity Pess.

Gali, J. and M. Gertler (1999) Inflation dynamics: a structural ecconometric analysis. *Journal of Monetary Ecconomics* 44, 195–222.

Gordon, R.J., ed. (1974) *Milton Friedman's Monetary Framework: A Debate with His Critics.* Chicago: University of Chicago Press.

―― (2009) *Macroeconomics,* 11th ed. New York: Pearson Education.

Keynes, J.M. (1936) *The General Theory of Employment, Interest, and Money.* London: Macmillan.

Laidler, D. (1995) Some aspects of monetarism circa 1970: a view from 1994, *Kredit und Kapital* 28, 323–45.

Lovel, M.C. (1986) Tests of the rational expectations hypothesis. *American Economic Review* 76, 110–24.

Maddala, G.S. (2001) *Introduction to Econometrics,* 2nd ed. Chichelster and New York: Wiley.

Nobay, A.R. and H.G. Johnson (1977) Monetarism: a historic-theoretic perspective. *Journal of Economic Literature* 15, 470–85.

Rotemberg, J.J. (1996) Prices, output, and hours: an empirical analysis based on a sticky price model. *Journal of Monetary Economics* 37, 505–33.

3
A Quest for the 'Missing Equations' in OECD Countries

3.1 Introduction

This chapter searches for the determinants of nominal income elasticities of total output and the price level in the postwar time series of some OECD countries. The elasticities indicate how annual changes in nominal GDP are divided between output changes and price level changes. This concept was described by Gordon (2009, ch 7) as one of unresolved questions in macroeconomics, since the time of Keynes (1936, chs 20, 21) and Friedman (1970, 1971). Friedman (1970) presented two macroeconomic models, one quantity theory and another income-expenditure theory, where the former system is closed by the assumption that total output is constant, and the latter by the assumption that the price level is constant. Friedman called either assumption a 'missing equation', but the term missing equation could be generalized to imply the equation involving intermediate situations between the two theories, which determines the proportions of output and price changes which accompany nominal income changes, because, as Friedman wrote in Gordon (1974, p. 45), 'the chief defect that this model [in Friedman (1971)] shares in common with the other two [that is the above two models] is that none of the three [models] has anything to say about the factors that determine the proportions in which a change in nominal income will, in the short-run, be divided between price change and output change ...'

This chapter focuses on postwar periods of eight OECD countries to inquire on what variables the nominal income elasticities depend in those countries; that is, we will deal with Australia, Canada, France,

Germany (until 1989, West Germany), Italy, Japan, the United Kingdom, and the United States. The periods covered are from around 1953 through 1998. From 1999 onward, many European countries joined the EU, so that some country-specific macro-variables are no longer available (particularly monetary aggregates), hence we limit the period of analysis up until 1998. We derive the variables (factors) determining nominal income elasticity of output *eo* (*eo* + *ep* = 1, where *ep* is nominal income elasticity of the price level) on the basis of firms' optimizing behavior on price change and their profits; that is, firms are assumed to carry out two-stage optimization.

The next section describes firms' behavior and, combining with it an equation explaining short-term economic growth of the economy, we derive the missing equation (nominal income elasticity of output) as a function of the variables which take on relatively exogenous characters. Section 3 estimates the equation for the period from around 1953 through 1998 using annual data from each of the above countries, and then compares the empirical results among them. Finally, Section 4 presents concluding remarks.

3.2   Firm behavior and factors generating nominal income elasticities of output *(eo)*

If nominal income, aggregate output, and the price level in some year are written as Y, Q, and P, respectively, one obviously has

$$Y = Q \cdot P.$$

If $g(z)$ means the growth rate of z, we find from the above that

$$g(Y) = g(Q) + g(P). \tag{3.1}$$

The symbols used here are collected after the concluding section, Section 4.

Next, writing $\Delta Y_t = Y_t - Y_{t-1}$, the nominal income elasticity of output *eo* can be written, omitting time subscripts, as

$$eo = \frac{\Delta Q}{\Delta Y} \frac{Y}{Q} = \frac{g(Q)}{g(Y)} = \frac{g(Q)}{g(P) + g(Q)} = \frac{1}{\frac{g(P)}{g(Q)} + 1}. \tag{3.2}$$

Hence, from (3.1), (3.2), and $ep = g(P)/g(Y)$, it follows that

$$eo + ep = 1.$$

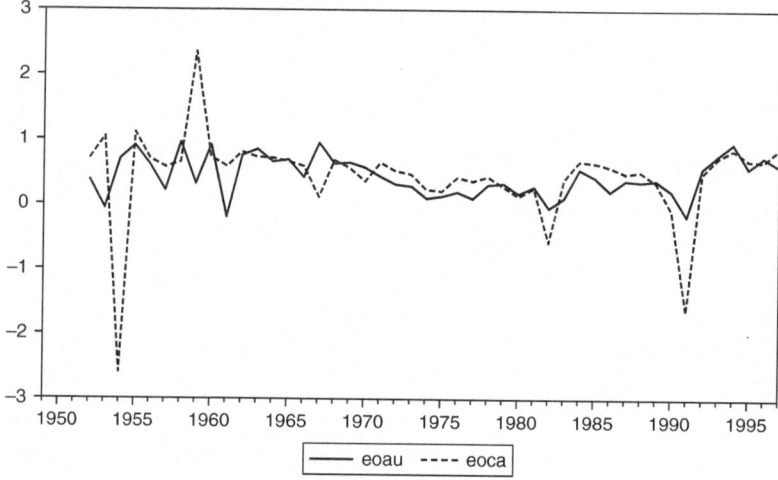

Figure 3.1 Elasticity of real outputs of Australia and Canada

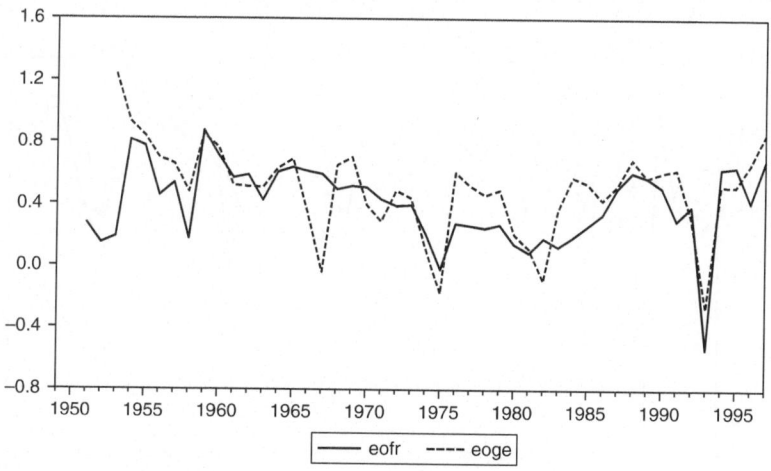

Figure 3.2 Elasticity of real outputs of France and Germany

Figures 3.1 through 3.4 show annual changes in *eo* for the eight countries we are concerned with. As (3.2) implies, since *eo* depends on $g(P)$ and $g(Q)$, we consider their determinants in this order.

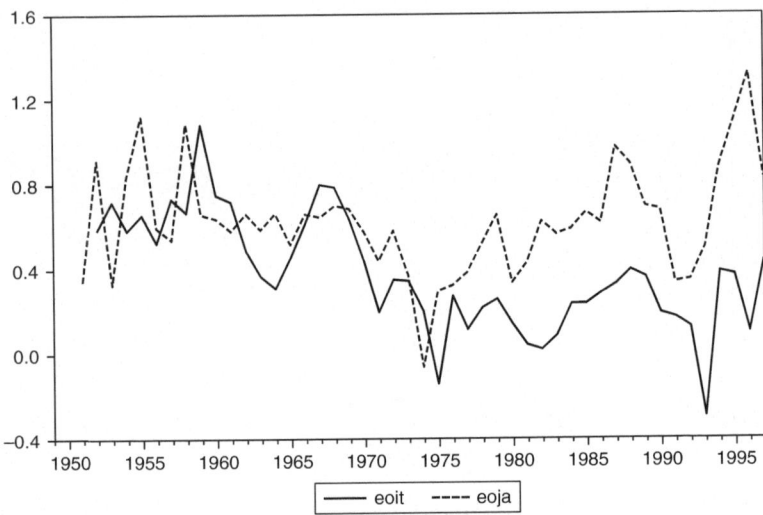

Figure 3.3   Elasticity of real outputs of Italy and Japan

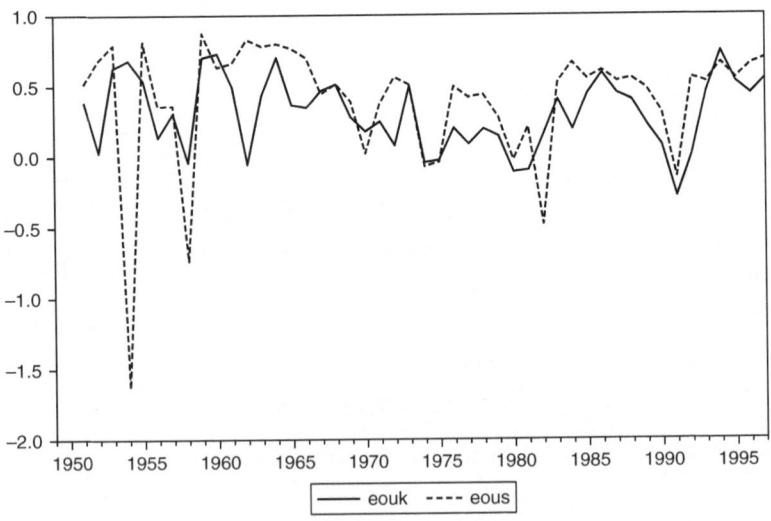

Figure 3.4   Elasticity of real outputs of the UK and the USA

3.2.1 The determinants of the inflation rate

We will obtain $g(P)$ on the basis of two-stage optimizing behavior of firms. Starting with a firm's short-run quadratic cost function (regarding price change), $C_t = C(s \geq t)$,[1] the derivation of the inflation equation is almost the same as in Chapter 1, hence we will reproduce the main components leading to the equation.

The firm's minimization of C_t results in, assuming discount factor δ equals unity,

$$p_t - p_t^* + k(gP_t - E_t gP_{t+1}) = 0. \tag{3.3}$$

We also have

$$p_t^* = mc_t + p^c{}_t + e_t, \tag{3.4}$$

where we assume that the ratio of the firm's optimal price at period t and the competitor's price p_t^c is equal to the log of the firm's marginal (production) cost relative to its trend level, mc_t,[2] and e_t is an iid (independent and identically distributed) error term with mean zero and a constant variance (that is, white noise; this assumption also applies to other error terms to appear in the following). Also, in (3.4),

$$mc_t = x_t + h(Q_t). \tag{3.5}$$

Here, x_t is a log of the nominal wage level relative to its trend, and $h(Q_t)$, a function of output Q_t, is a log of a reciprocal of marginal productivity of labor relative to its trend level. (See note 2 again; the assumption that the firms facing a monopolistically competitive product market might require that $mc_t = p_t^* - p^c{}_t - e_t$ (from (3.4)) be augmented by $\ln(1-\gamma)$ where γ is a reciprocal of the price elasticity of product demand; however, as long as γ remains constant, ignoring this factor will not cause any problem.) Also, we assume that $h(Q_t)$ will increase in the short-run when some components of aggregate demand increase, while $h(Q_t)$ will decrease when technical progress occurs. In the former case, output increases at least temporarily, which will result in increases in the reciprocal of the marginal productivity of labor (on assuming the decreasing marginal productivity of labor in the short-run; that is, $h'(Q_t) > 0$), but in the case of technical progress, $h(Q_t)$ will decrease with the same level of output.

Next, from (3.3), (3.4), and (3.5), one obtains

$$gP_t = E_t gP_{t+1} + \frac{1}{k}(p_t^* - p_t) = E_t gP_{t+1} + \frac{1}{k}[x_t + h(Q_t) + p^c{}_t + e_t - p_t].$$

When the above relationship is applied to the macro-economy, it would be natural to suppose that the firm's actual price is approximately equal to the competitors' prices, that is, $p_t = p_t^c$. Then the above equation becomes

$$gP_t = E_t gP_{t+1} + \frac{1}{k}[x_t + h(Q_t) + e_t],$$

where k is a positive constant. Changing the notation slightly such that $x_t/k = w_t$, $h(Q_t)/k = l_t$, and $e_t/k = u_t$, the above relationship becomes

$$gP_t = E_t gP_{t+1} + w_t + l_t + u_t, \tag{3.6}$$

where l_t is a function of some aggregate demand components and/or of the 'state of technology'; the higher value of demands will raise the price at least in the short-run, while technical progress will lower it without affecting output levels.

Finally, we assume that the log of an actual wage level relative to its trend, x_t, is a positive function of some measure of excess demand for labor (see note 2 again).[3] We represent this measure by the rate of change of the unemployment of labor, gur_t.[4] Hence,

$$w_t = a_0 + a_1 gur_t + v_t, \tag{3.7}$$

where a_1 is a positive constant and v_t is an iid error term with a zero mean.[5]

Substituting (3.7) into (3.6), and attaching a positive coefficient on $E_t gP_{t+1}$ and l_t, we find

$$gP_t = a_0 + a_1 E_t gP_{t+1} + a_2 gur_s + a_3 l_s + \eta_t, \tag{3.8}$$

where the subscript s implies either t (the current year) or $t - j$ ($j > 1$: some past year); the following regression will show which subscript best describes the actual economic process. Also, $\eta_t \equiv u_t + v_t$ is a zero mean iid error term.

3.2.2   The determinants of output growth

We will now look for the determinants of output growth in year t as a function of past and contemporaneous macroeconomic variables. Since the explanatory variables are mainly short-run determinants, they consist mostly of demand factors as well as supply factors.

As demand factors, we consider (a) the central government's budget deficits (in ratio form, expenditure/revenue, written as er), (b) current account surpluses (in ratio form, export/import, written as em), and (c) money supply ($m2$) that includes cash in circulation as well as demand and time deposits. Also, the supply factors in our growth equation are (d) the total factor productivity (tf) and (e) the unemployment rate for labor. The first two factors are written in ratios, which can dispense with procedures to get constant-price values. The *levels* of those five factors, (a) through (e), affect the output *level* by way of its demand and supply sides; hence it would be their *growth* rates that affect the *growth* of output (gQ).

If the annual growth rate of er is written as ger, and so on, then the output growth would be written as

$$gQ_t = b_1 ger_s + b_2 gem_s + b_3 gm2_s + b_4 gtf_s + b_5 gur_s + \mu_t, \qquad (3.9)$$

where μ_t is an iid error term with a zero mean, and subscript s in (3.9) implies either t (the current year) or $t-j$ ($j > 1$).[6]

Substituting the determinants of gP and gQ, (3.8) and (3.9), into the definition of eo (3.1), while observing note 6, and then linearizing the resulting expression, one obtains an equation for eo as a function of explanatory variables, which is to be estimated in the next section:

$$eo_t = c_0 + c_1 ger_s + c_2 gem_s + c_3 gm2_s + c_4 gur_s + c_5 gtf_s + c_6 gPe_{t+1} + \varepsilon_t \qquad (3.10)$$

where $gPe_{t+1} \equiv E_t gP_{t+1}$, which is an expected inflation rate for year $t+1$ formed at year t. The terms c_k ($k = 0$ through 6) are the parameters to be estimated. Also, ε_t is a zero mean iid error term.

3.3   Estimation of nominal aggregate income elasticity of output *(eo)*

Let us begin with expectation formation on inflation, gPe_{t+1}. The rational expectations hypothesis implies

$$gP_{t+1} = gPe_{t+1} + v_{t+1} \qquad (3.11)$$

where gP_{t+1} is the actual inflation rate, v_{t+1} is an iid error term with mean zero, and the two terms on the right-hand side are not correlated (that is, $\text{cov}(gPe_{t+1}, v_{t+1}) = 0$). See, for example, Lovell (1986).

Substituting gPe_{t+1} from (3.11) into (3.10) yields

$$eo_t = f(s \leq t) + c_6(gP_{t+1} - v_{t+1}) + \varepsilon_t, \tag{3.10'}$$

where $f(s \leq t)$ represent the first six terms on the right-hand side of (3.10). If the ordinary least squares method is applied to the above, the resulting estimates are not consistent because gP_{t+1} and v_{t+1} are correlated in view of (3.11). Hence one needs to estimate gP_{t+1} in (3.10') by the instrumental variable method. Actually, we estimate gP_t on the instruments of past gPs, because in period t, agents cannot know the values gP_{t+1}; and then we shift estimated gP_t one period forward to make it gPe_{t+1}. Instruments were then chosen to be gP_{t-1}, gP_{t-2}, gP_{t-3}, for France, Germany, and the United Kingdom; while for Australia, Canada, Italy, Japan, and the USA, the instruments were reduced to the past two gPs. These instrument choices were based on higher adjusted coefficients of determination and better LM statistics (for residual autocorrelation) than the other cases [7,8] The estimated expected inflation rate for $t+1$, $gPe_t(1)$, was then used in (10') in place of gP_{t+1}, and we next estimated (3.10).[9,10]

To obtain the growth of total factor productivity, gtf, consider the following Cobb-Douglas production function of the economy:

$$Q = (tf)K^{0.3}N^{0.7},$$

where tf is the level of total factor productivity, K is the aggregate existing stock of capital, and N is the number of labor employed, all in year t (t is omitted for brevity of notation). Regarding Germany, France, the UK, and the USA for most of the years covered here, we used the real gross capital stock estimates of O'Mahony (1996); other countries' estimation methods and data sources are described in the Appendix A.

We also suppose that the capital's share in GDP (the elasticity of output regarding capital in use) is 0.3, and the labor's share (the elasticity of output regarding labor is 0.7.[11] Hence the constant returns to scale relative to the two factors are assumed. Taking the logarithm of the production function and differentiating it with respect to time yields

$$gtf = gQ - 0.3gK - 0.7gN,$$

where gtf is the growth rate of tf, and similarly for other terms beginning with g.

The whole period of estimation is from around 1953 through 1998 (France and Japan start with 1954 and 1952, respectively; the effective last year is 1997 because expectational variables take their one year forward values). Japan has seen consecutive deflationary years in the GDP deflator since 1999. Hence Japan's *eo* showed erratic behavior during that time (for example, it was 2.845 in 2000, and −136 in 2003; *eo* and *ep* normally take on values within $0 \leq eo, ep \leq 1$). Also, the starting of the EU made it hard to prepare monetary aggregates of each EU member country measured by its previous monetary unit from 1999 onward. Therefore, 1998 was set as the last year for the eight countries. The data are aggregated annually. The data sources for this chapter are described in Appendix B.

We now turn to each country's estimation. A result for Australia, which comes first in the alphabetical order of the countries we examine, is shown below, where the variables in (3.10) are suffixed by '*au*' to remind us that they are 'Australian' variables. Also, to simplify notation, $gPe_{t+1}au$ is written as *gpeau*(1), and similarly for other countries. The estimation turned out to be

eoau	*gemau*	*gerau*	*gmau*(−1)	*gurau*	*gtfau*	*gpeau* (1)
aver.	−0.553	−0.169	−1.115	−0.241	3.193	−4.221
=0.437	(0.010)	(0.751)	(0.071)	(0.008)	(0.069)	(0.001)

(1953–97) $\overline{r^2} = 0.660$, $SER = 0.180$, $DW = 2.376$, $LMP = 0.327$, $BPGP = 0.067$, $WP = 0.291$, where the intercept is omitted from the table.

The first line below the variable shows the coefficient values, and figures in parentheses are the *p*-values (significance levels in *t*-distribution) for the null of the coefficients being zero. The *p* indicates the probability of committing an error when rejecting the null hypothesis. The 'aver.' below *eoau* is its sample mean (for 1% increase in nominal GDP this country saw, on average, about 0.44% annual increase in real GDP, and about 0.56% rise in inflation). The lag length of zero or *j* ($j = 1$ or 2) was tried for each of the first five explanatory variables, and the one with the lowest *p* was chosen. The period in parentheses is the effective period of estimation, where 1998 is lost because expected inflation uses the value one year ahead. The term $\overline{r^2}$ is the adjusted coefficient of determination, *SER* is the standard error of regression, and *DW* is the Durbin-Watson ratio. This *DW*-ratio shows that the errors have no autocorrelation ($d_U = 1.859$, above which one cannot reject the null of there being no autocorrelation). Also, the *p*-value for the Breush-Godrey LM test (*LMP*) is large enough not to reject the null that errors do not have serial correlation. As for heteroskedasticity of error terms, the *p*-value for the Breush-Pagan-Godfrey test (*BPGP*) cannot reject the null of no heteroskedasticity at 5% significance level.[12]

Besides, the *p*-value for the White heteroskedasticity test (*WP*) indicates the error terms have no heteroskedasticity. For this country, broad money *m2* consists of money and quasi-money.

The first point to note would be that the larger export/import ratio lowers *eo*, and raises *ep* (*eo* + *ep* = 1). Increased money supply lowers output growth compared to inflation in Australia (see the rightmost expression for *eo* in Equation (3.2)). Higher growth of the unemployment rate lowers *eo*, but a higher rate of technical progress raises it, though the latter is significant only at the 7% level. Higher inflation expectation lowers *eo*, so that it raises *ep* as the theory predicts.

Estimation for Canada yielded the following result:

	eoca	gemca	gerca(−1)	gmca	gurca(−1)	gtfca	gpeca(1)	dca	dcaa
aver.	−1.206	−1.349	−0.271	−0.583		8.695	−4.514	−2.424	1.694
= 0.441	(0.023)	(0.054)	(0.719)	(0.010)		(0.000)	(0.015)	(0.000)	(0.000)

(1953–97) $\overline{r^2}$ = 0.947,   *SER* = 0.160,   *DW* = 1.893,   *LMP* = 0.955, *BPGP* = 0.000, *WP* = 0.014, where the first dummy variable *dca* takes on 1 for 1954 and 1991 and 0 otherwise, while the second one *dcaa* equals 1 in 1959 and 0 otherwise (see Figure 1).

The monetary aggregate is the sum of money and quasi-money. The *p*-values in both the Breush-Pagan-Godfrey and the White heteroskedasticity tests show its existence, hence we use the White heteroskedasticity consistent covariance matrix (*p*-values for the coefficients are the results of this).

Note that, while in Australia, increases in broad money supply have an inflationary effect (though it is significant only at 8% level), Canada does not show significant effects on either *eo* or *ep*. This might be the result of enhancing effects of money on both output and inflation, which made the effects on both *eo* and *ep* insignificant (that is, the effects of money on output and inflation canceled each other out).

Turning now to France, the corresponding result is as follows:

eofr	gemfr(−1)	gerfr	gmfr	gurfr(−1)	gtffr	gpefr(1)	dfr
aver.	−0.076	−0.677	0.658	0.055	4.042	−6.786	−0.992
= 0.418	(0.690)	(0.016)	(0.077)	(0.461)	(0.002)	(0.000)	(0.000)

$(1954\text{--}97)$ $\overline{r^2} = 0.926$, $SER = 0.070$, $DW = 1.378$, $LMP = 0.478$, $BPGP = 0.004$, $WP = 0.027$. Dummy variable dfr takes on 1 for 1993, but 0 in other years. The above DW-ratio lies in the range where one cannot judge if serial correlation exists, but LMP clearly shows there is no serial correlation. Also, in view of $BPGP$ and WP, we used the White heteroskedasticity consistent covariance matrix.

One notes that, contrary to Australia, higher money growth leads to higher output growth rather than higher inflation, where money is broad money, $m2$.

The fourth country, Germany, exhibits the opposite effect of export/import ratio to Australia and Canada, and the whole result turns out as

eoge	gemge	gerge	gmge	gurge	gtfge	gpege (1)
aver.	1.174	−0.275	0.494	−0.167	7.634	−9.096
= 0.467	(0.026)	(0.614)	(0.363)	(0.003)	(0.000)	(0.000)

$(1953\text{--}97)$ $\overline{r^2} = 0.687$, $SER = 0.173$, $DW = 1.782$, $LMP = 0.345$, $BPGP = 0.801$, $WP = 0.808$.

The increasing export/import ratio in Germany raises eo (and lowers ep); the government expenditure/revenue ratio accompanies an insignificant coefficient. In this country, increasing growth of $m2$ did not have significant effect on eo or ep.[13]

Estimation for Italy can be summarized as follows:

eoit	gemit	gerit	gmit	gurit	gtfit	gpeit(1)
aver.	−0.075	−0.712	1.734	−0.692	4.346	3.329
= 0.372	(0.773)	(0.026)	(0.000)	(0.000)	(0.001)	(0.000)

$(1953\text{--}97)$ $\overline{r^2} = 0.785$, $SER = 0.127$, $DW = 1.600$ (this figure does not ensure the errors have no autocorrelation), $LMP = 0.487$, $BPGP = 0.299$, $WP = 0.026$.

Although $BPGP$ is large enough to preclude heteroskedasticity, WP is not. Hence we use the White heteroskedasticity consistent covariance matrix. Italy's increase in government expenditure/revenue ratio reduces real growth relative to inflation, which is the same as the previous estimations for Canada and France. Italy's monetary aggregate is measured by *m2*.

The Japanese postwar experience yields the following results:

eoja	*gemja*	*gerja*	*gmja*(−1)	*gurja*	*gtfja*	*gpeja*(1)	*dja*
aver.	−0.374	0.998	−1.282	0.366	3.801	−5.191	−0.237
= 0.628	(0.052)	(0.130)	(0.005)	(0.038)	(0.000)	(0.000)	(0.000)

(1952–97) $\overline{r}^2$ = 0.748, SER = 0.128, DW = 1.747, LMP = 0.572, BPGP = 0.061, WP = 0.288.

Here, a dummy *dja* takes on 1 for 1974–83 and 1991–93, while it is 0 for other years. The monetary aggregate used for Japan is *m2*. Although the Breush-Pagan-Godfrey heteroskedasticity test is significant only at the 7% level, the *p* for White's test is large enough; hence we adopt the above results. (Use of White's matrix yields essentially similar results here.) Similarly to Australia and Canada, an increasing ratio of export/import makes for higher inflation rather than higher growth, but the effect is in opposite direction to Germany. Particularly notable for Japan is that higher growth in unemployment makes for higher *eo*, contrary to negative significant effects in other countries except for France, where the effect is not significant. This growth-enhancing effect of higher unemployment rates would be due to its exceptionally low unemployment rates (during the estimation period), so that higher unemployment made more workers available for firms; in other words, higher unemployment in Japan worked on its economy as a favorable supply-side effect. The sample averages of the unemployment rate for 1952–98 of Japan was 2.1%, while the US and the UK counterparts were 5.8% and 5.0%, respectively.

Estimation for the UK is as follows:

eouk	*gemuk*	*geruk*	*gmuk*(−1)	*guruk*	*gtfuk*	*gpeuk*(1)	*duk*
aver.	−0.513	−0.736	−0.746	−0.516	4.802	−1.281	−0.370
= 0.304	(0.221)	(0.032)	(0.017)	(0.000)	(0.000)	(0.020)	(0.005)

(1952–97) $\overline{r}^2$ = 0.796, SER = 0.118, DW = 1.855, LMP = 0.116, BPGP = 0.150, WP = 0.846.

A dummy variable *duk* takes on 1 for 1991 and 0 otherwise. It is noted that the two demand factors (the government expenditure/revenue ratio and broad money supply (*m*3)) make for higher inflation relative to higher real growth.

Finally, the relationship for the USA was estimated as

eous	*gemus*(−1)	*gerus*	*gmus*	*gurus*(−1)	*gtfus*	*gpeus*(1)	*dus*
aver.	0.473	−0.644	0.347	−0.248	14.501	−3.628	−1.355
= 0.400	(0.086)	(0.023)	(0.624)	(0.001)	(0.000)	(0.018)	(0.000)

(1952–97) $\overline{r^2}$ = 0.894, *SER* = 0.145, *DW* = 2.079, *LMP* = 0.764, *BPGP* = 0.000, *WP* = 0.002. The *BGP* being insignificant, the above estimation uses the Newey-West heteroskedasticity consistent covariance matrix (the White matrix yielded similar results but some *p*-values were larger). The dummy *dus* takes on 1 for 1954 and 1958, and 0 for other years.

In the US outcome, the growth of government expenditure/revenue ratio made for lower *eo*, as in most other countries. Also, as in Canada and Germany, an increasing *m*2 growth rate did not have significant effects on *eo* or *ep*; this might be due to positive effects of higher *m*2 on both real output and inflation, so that higher *m*2 did not result in significant changes in *eo* or *ep*.

Although the results for the UK and Japan are quite similar to those described in Chapter 1 (even though Japan uses different measures for labor market tightness), the US's are not; the effect of technical progress appears here to be positive and significant, while in Chapter 1 it appeared insignificant. The similarity was not gained even when, in Chapter 1, a dummy was used (taking on 1 for 1954 and 1958) and/or *gtfus* was used, which is derived on assuming constant income shares, that is, 0.7 for labor and 0.3 for capital. For now, therefore, reconciling the two results has to be left for future work.

The above examination of individual countries may reveal some useful comparative features when the estimated coefficients are shown together:

The first point one may note from the table is that, in five countries, the government expenditure/revenue ratio makes for lower real growth relative to inflation, which is in accordance with much regression work, such as Barro and Sala-i-Martin (2004, ch 12) who showed that government consumption tends to check per capita income growth. The second point is that larger export/import ratios and money growth rates do not necessarily spur real growth relative

Table 3.1 Coefficient comparison among eight countries

	gem	ger	gm	gur	gtf	gpe (1)	**aver.**
Australia	−0.553***	?	−1.115*	−0.241***	3.193***	−4.221***	0.437
(1953–97)							(4) [3]
Canada	−1.206**	−1.349*	?	−0.583***	8.695***	−4.514***	0.441
(1953–97)							(3) [2]
France	?	−0.677**	0.658*	?	4.042**	−6.786***	0.418
(1954–97)							(5) [6]
Germany	1.174**	?	?	−0.167***	7.634***	−9.096***	0.467
(1953–97)							(2)[4]
Italy	?	−0.712***	0.734***	−0.692***	4.346***	−3.329***	0.372
(1953–97)							(7) [5]
Japan	−0.374*	?	−1.282***	0.366**	3.801***	−5.191***	0.628
(1952–97)							(1)[1]
UK	?	−0.736**	−0.746**	−0.516***	4.802***	−1.281**	0.304
(1952–97)							(8) [8]
USA	0.473*	−0.644***	?	−0.248***	14.501***	−3.628**	0.400
(1952–97)							(6) [7]

Notes: (1) The numbers in the cells are the coefficients of the variables, with the country attributes omitted. (2) The periods below the country names are effective estimation periods. (3) *, **, ***: significant at 10%, 5%, and 1% levels, respectively; ?: insignificant at 10% level. (4) The figures in the last column are the sample average of *eos*. (5) The figures below the average *eos* in parentheses show the descending order of their size. (6) The figures in square brackets are the descending order of the average growth rate of real GDP.

to inflation. That may be due to these two variables affecting both growth and inflation positively and, as the last relation in (3.2) (that is, $eo = 1 / \left[\frac{g(P)}{g(Q)} + 1 \right]$) implies, the changes in $g(P)$ and $g(Q)$ in the same direction may well make the coefficients on the export/import ratio or the money growth rate insignificant; or the results may turn out to be country-specific.

Thirdly, one notes that technical progress and inflation expectations one year ahead have unanimous effects on *eos*; and most coefficients are significant at the 1% level, except for Australia's *gtf* and UK and the US's inflation expectations.

Finally, for years 1952 through 1998, the order correlation between *eos* and average growth rates, which is computed from the last column's two orders, is 0.738, which is quite high. Also, the order correlation between each country's average *ep* and average rate of inflation is 0.667. Although the inflation data are not shown, the order correlation between average growth rates and average inflation rates is -0.095, which is very low, but as the sample number increases, this result might change.

3.4 Conclusions

This chapter has been concerned with the economic factors that affect the division of annual changes of nominal GDP between changes in real output and the price level; see Nobay and Johnson (1977) who maintained that the description of output-inflation dynamics in the wake of nominal disturbances was the main subject of classical (pre-Keynesian) monetary economics. See also Laidler's (1995) historical description on the developments of monetarist economics, where the author notes Friedman's concern about the fact that the proportions of output and price changes following (or occurring concomitantly with) nominal income changes have not been explained so far (Laidler 1995, p. 338, footnote 16).

This chapter might be viewed as an initial attempt at addressing the above state of affairs in macro- and monetary-dynamics.

List of Symbols

eo	nominal income elasticity of output.
ep	nominal income elasticity of the price level.
C_t	price adjustment cost in period t.
g	growth rate operator; for example, $g(z)$: growth rate of z.
ur	rate of unemployment.
er	government's expenditure/revenue ratio.
em	export/import ratio.
m2	money supply consisting of cash currency and demand and time deposits.
tf	level of total factor productivity.
$h(Q)$	log of a reciprocal of marginal productivity of labor relative to its trend, as a function of output Q.
l	vector, or a subset of (*ger gem gm2 gtf*).
LMP	*p*-value in the Breush-Godfrey LM test for serial correlation.
BPGP	*p*-value in the Breush-Pagan-Godfrey test for heteroskedasticity.
WP	*p*-value in the White heteroskedasticity test.

Appendix A. Estimation of Capital Stock based on the Perpetual Inventory Method

First, for Australia, Canada, and Italy, on which capital stock estimates are not available, we describe the estimation method based on the perpetual inventory method $K_i = 0.92K_{i-1} + I_i$, where i refers to some year, K_i the real gross capital stock at the start of year i, and I_i is real gross investment made during year i.

The annual physical depreciation rate is assumed as 8% which seems considerably standard (see, for example, Benhabib and Spiegel (2000), and Luintel and Kahn (1999)).

Taking Canada for example, if $g(K_1)$ means the growth of K_1, we have

$$g(K_1) = (K_1 - K_0)/K_0 = (I_1 - 0.08K_0)/K_0 = (I_1/K_0) - 0.08$$

(using the above formula)

so that, generally,

$$g(K_i) = (K_i - K_{i-1})/K_{i-1} = (I_i - 0.08K_{i-1})/K_{i-1} = (I_i/K_{i-1}) - 0.08.$$

Note that here I_i is constant-price gross fixed capital formation.

If we now assume $g(K_0) = g(K_1) = \cdots = g(K_5)$, from the above equation (that is, from $g(K_i) = (I_i/K_{i-1}) - 0.08$ $(i = 1, \ldots, 5)$), we have, as for real fixed capital formation, that $g(K_0) = g(I_1) = 0.1112$, where, for Canada, 0 refers to 1951, $g(K_1) = g(I_2) = 0.1260$ (for Canada, nominal gross fixed capital formation in $1951 = 4.42$, and it is 5.10 in 1952, while the GDP deflator [nominal gross domestic product/gross national product, 1975 prices; see *International Financial Statistics*, 1979, p.130] for $1951 = 10.154$, and it is 11.283 for 1952; from those figures, $g(I_1)$ is calculated as above), and in a similar way, we find $g(K_3) = 0.1373$, $g(K_4) = 0.2020$, $g(K_5) = 0.0672$.

Next, we average the above six $g(K_i)$s to find $av. \equiv 0.6237/6 = 0.1040$. This is a common $g(K_i)$ assumed above; the number of years, six, for getting $av.$ has no special meaning, which seems to be neither too large nor too small.

Then, using again the formula, $K_0 = K_{1951} = I_1$ (that is, $I_{1952}/(av. + 0.08) = 61.32$, $K_1 = K_{1952} = I_2$ (that is, $I_{1953}/(av. + 0.08) = 69.5$, and the Ks of the following years are calculated in a similar way.

Appendix B. Data Sources

The following series are drawn from *International Financial Statistics* (*IFS*), International Monetary Fund: nominal and real GDP, GDP deflators, export/ import ratios, government expenditure/revenue ratios, $m2$ or money plus quasi-money, and gross fixed capital formation.

For Australia, Canada, and Italy, gross capital stocks were estimated in the way described in Appendix A, where nominal gross capital formations were made constant-price counterparts using GDP deflators of each country.

For France, Germany, Japan, the USA, and the UK, we used the estimates on gross capital stocks by O'Mahony (1996) for years 1952 through 1989, while for later years we estimated them using the formula in the perpetual inventory method, also using constant-price capital formation data adapted from *IFS*.

Unemployment rates of all countries were drawn from *IFS* in and after 1985. The counterparts for 1952 through 1985 were derived from *Foreign Statistics Annuals* of the Bank of Japan.

(Original sources of unemployment rate statistics are: *Monthly Summary of Statistics* for Australia; *Bank of Canada Review* and *Canadian Statistical Review* for Canada; *Bulletin Mensuel de Statistique* and *UN Monthly Bulletin of Statistics* for France; *Wirtschaft und Statistik, Statistisches Jahrbush, Statistischer Wochendienst*, and *Monthly Report of the Deutsche Bundesbank* for Germany; *UN Monthly Bulletin of Statistics*, and *Bollettino Mensile di Statistica* for Italy; *Labor Research Report* for Japan; *Monthly Digest of Statistics* for the UK; and *Survey of Current Business* and *Monthly Labor Review* for the USA.)

Notes

1. Let us note that the above C_t concerns the cost involved in price changes, and is different from the cost involved in the firm's output production; see later discussion.
2. See Gali and Gertler (1999, p. 199 et seq.) or Gali (2008, for example, p. 18) for similar assumptions: the above paper considers percentage deviations of firms' marginal cost from the steady state value and of aggregate output from its natural level, both as determinants of price inflation.
3. With w_t being the gap between the actual wage level and its trend level, it is a variable conceptually similar to the rate of change of the wage level.
4. As our estimation shows below, the rate of change of the unemployment rate explains the behavior of w_t better than the level of unemployment

does. One reason for this would be that the wage level is taken as the ratio to its trend level.

5. One might consider x_t to depend on $E_t g P_{t+1}$, whose coefficient would be unity according to the natural rate hypothesis. However, on a short-run horizon, the coefficient of this expectation term may well be smaller than unity. In any case, the expectation term also appears as a first term in (3.5). Hence, neglecting the term in (3.7) does not cause any problem in the following discussion.

6. An explanatory variable l_s in (3.8) is actually a vector, or a subset of (ger_s gem_s $gm2_s$ gtf_s).

7. However, equalizing the lag number to either two or three for those countries yielded similar results.

8. We also tried estimation using, as instruments, lagged values of $gm2$ as well as gP. These extra instruments, however, yielded lower adjusted coefficients of determination and/or worse LM statistics.

9. This means that we use the two-stage least squares method.

10. Alternatively, using another definition of rational expectations, $gPe_{t+1} = E_t(gP_{t+1}|I_t)$, where I_t is the information available at t, one may regard the estimated value of the above autoregression, forwarded one period ahead, as gPe_{t+1} and proceed in the same way as in the text. See Maddala (2001, pp. 419–22).

11. These values are also assumed in Benhabib and Spiegel (2000). Assuming the two elasticities to be constant over time and cross-sectionally is considered a standard procedure.

12. Use of the White heteroskedasticity consistent covariance matrix yielded similar coefficients and diagnostic figures.

13. For Germany, introducing a dummy variable taking on 1 in 1993 improves the p-values of *gemge* and *gurge* only slightly, but does not change the result, hence we proceed without using it.

References

Barro, R.J. and X. Sala-i-Martin (2004) *Economic growth*, 2nd ed. Cambridge, Mass.: MIT Press.

Benhabib, J. and M.M. Spiegel (2000) The role of financial development in growth and investment. *Journal of Economic Growth* 5, 341–60.

Calvo, G.A. (1983) Staggered prices in a utility maximizing framework. *Journal of Monetary Economics* 12, 383–98.

EViews 6 (2007) *User's Guide*. Irvine, CA: Quantitative Micro Software.

Friedman, M. (1970) A theoretical framework for monetary analysis. *Journal of Political Economy* 78, 193–238.

—— (1971) A monetary theory of nominal income. *Journal of Political Economy* 79, 323–37.

Gali, J. (2008) *Monetary Policy, Inflation, and the Business Cycle*. Princeton: Princeton University Press.

Gali, J. and M. Gertler (1999) Inflation dynamics: a structural econometric analysis. *Journal of Monetary Economics* 44, 195–22.

Gordon, R.J., ed. (1974) *Milton Friedman's Monetary Framework*. Chicago: University of Chicago Press.

—— (2009) *Macroeconomics*, 11th ed. New York: Pearson Education.

Keynes, J.M. (1936) *The General Theory of Employment, Interest, and Money*. London: Macmillan.

Laidler, D. (1995) Some aspects of monetarism circa 1970: a view from 1994. *Kredit and Kapital* 28, 323–45.

Lovell, M.C. (1986) Tests of the rational expectations hypothesis. *American Economic Review* 76, 110–24.

Luintel, K.B. and M. Kahn (1999) A quantitative reassessment of finance-growth nexus: evidence from a multivariate VAR. *Journal of Development Economics* 60, 381–405.

Maddala, G.S. (2001) *Introduction to Econometrics*, 3rd ed. Chichester and New York: J. Wiley and Sons.

O'Mahony, M. (1996) Measures of fixed capital stocks in the post-war period: a five-country study. In B. van Ark et al., eds. *Quantitative Aspects of Post-War European Economic Growth*, pp. 165–214. Cambridge: Cambridge University Press,

Nobay, A.R. and H.G. Johnson (1977) Monetarism: a historic-theoretic perspective. *Journal of Economic Literature* 15, 470–85.

Rotemberg, J.J. (1996) Prices, output, and hours: an empirical analysis based on a sticky price model. *Journal of Monetary Economics* 37, 505–33.

Part II

4
The NAIRU, Potential Output, and the Kalman Filter: A Survey and Method of Estimation

4.1 Introduction

The concepts of NAIRU (non-accelerating inflation rate of unemployment) and potential output have received increasing attention recently as the rate of inflation has occupied a higher position in monetary policy discussion. This seems particularly true as more central banks of various economies regard inflation as a main targeting variable in their policy operations (Bernanke and Mishkin (1997)). For, if the central bank can correctly forecast the future course of the NAIRU and if it can steer the economy to that position, then the policy authority can obtain a stable rate of inflation. In addition, if it can control the growth of money supply, at least in the medium-term horizon, the rate of inflation of the economy would be controlled at a low and stable level.

In this chapter we first survey the work dealing with the NAIRU and potential output, and try to suggest several points to be considered in future work. The second part of the chapter estimates the NAIRU and potential output of Japan, 1950–2002, using the Kalman filter. The Kalman filters, or more generally state-space models, were originally developed by control engineer R. Kalman to express dynamic systems involving unobservable as well as observable variables, such as the NAIRU or potential output, and to estimate those variables. The main component of the dynamic systems to be developed below is a number of variants of the expectation-augmented Phillips curve. We also examine the correspondence between excess demands in the labor and product markets, on the one hand, and the turning points

59

of postwar Japan's business cycles (inventory cycles), on the other hand.

This chapter is organized as follows: The next section surveys the literature on the NAIRU (or the natural rate of unemployment) and potential output. Section 3 estimates those unobservable variables, which are obviously important for policymaking as will be clear from the definition of the NAIRU. Section 4 concludes with several remarks.

4.2   From a policy menu to the NAIRU and potential output: a survey

The Phillips curve, a relationship between the rate of inflation of the price level or the wage rate on the one hand, and the rate of unemployment on the other, has occupied an important position in policy discussion. When the existence of this relationship was first claimed by Phillips (1958), it was thought to provide a policy menu for monetary and fiscal authorities (Samuelson and Solow 1960).[1] It was Phelps (1967) and Friedman (1968), however, who later denied the stable negative relationship between the rate of price change and the rate of unemployment. They then showed that when the changes in wage rates (prices) are made in real terms, that is, the courses of those changes in the next period are correctly anticipated, the negative tradeoff between the rate of wage (price) changes and the unemployment rate would disappear. Friedman called the unemployment rate which prevails in those expectational equilibria the 'natural rate of unemployment'. He describes this rate as:

> the level that would be ground out by the Walrasian system of general equilibrium equations, provided there is embedded in them the actual characteristics of the labor and commodity markets, including market imperfections, stochastic variability in demands and supplies, the cost of gathering information about job vacancies and labor availabilities, the cost of mobility, and so on. (Friedman 1968, sec. I)

Hence, when an economy that finds itself in an expectational equilibrium is viewed in a static way as in a snap shot, the Phillips curve would have a shape of a vertical line which passes through a point on a horizontal unemployment rate line. This point is exactly the natural rate. When the economy is located on this point, expectations on the rates of change of wages as well as prices are realized and those rates of change would be stable, because the economy has equilibrium markets with appropriate 'slackness' as mentioned in Friedman's remarks above. The second characteristic of the natural rate prompted economists to call it the NAIRU (non-accelerating inflation rate of unemployment, or, more appropriately, non-changing inflation rate of unemployment).

We noted that the above description of the NAIRU is relevant in a snap shot of the economy at a point in time. When the economy is viewed as evolving over time, however, the various factors affecting the slackness in market equilibria would change with a lapse of time; that is, the natural rate of the economy would change from quarter to quarter or from year to year. This point was noted earlier by R.A. Gordon (1977), who argued that the natural rate is not on a vertical line if it is seen over some interval of time.

Although there have been some contributions meanwhile, it was as late as the work of R.J. Gordon (1997) that the argument of R.A. Gordon (1977) was implemented into empirical analysis. R.J. Gordon (1997) is, as far as we know, the first work that tried to estimate the NAIRU using the Kalman filter, which is a variant of state-space models or time-varying-parameter models (Kim and Nelson (1999), Hamilton (1994), and Harvey (1993), develop extensive discussions of state-space models and the Kalman filter). Gordon's (1997) model specifically is composed of

$$\pi_t = a(L)\pi_{t-1} + b(L)(u_t - u^N_t) + c(L)z_t + e_t,$$

$$u^N_t = u^N_{t-1} + \varepsilon_t, \tag{4.1}$$

where π_t is the rate of inflation, u_t is the actual unemployment rate in period t, u^N_t is the natural unemployment rate in period t, $a(L)$, $b(L)$, and $c(L)$ are polynomials (some linear functions) in the lag operator L (that is, $L\pi_t = \pi_{t-1}$ and $L^2\pi_t = \pi_{t-2}$, and so on), z_t is a vector of supply shock variables, potentially undergoing sudden changes, such as petroleum prices and international commodity prices. Also, e_t and ε_t are serially uncorrelated error terms and are uncorrelated with each other. In the above, u^N_t is assumed to follow a random walk.[2] Gordon derived several alternative paths of u^N_t for postwar US economy on applying alternative standard deviations for ε_t, which are given exogenously.

Staiger, Stock, and Watson (1997a, b) estimate the time-varying NAIRU using what is called Fieller's method of constructing a confidence interval for the ratio of the means of two dependent normal random variables. Their conclusion from their estimation attempt, however, is not a very encouraging one: that their NAIRU estimates are not precise enough for forecasting long-run inflation. Though King, Stock, and Watson's (1995) formulation of the Phillips curve is different from those in the above two papers, the conclusion obtained here has a feature common to them, which means that the prediction of future inflation based on the Phillips curve using the NAIRU is too imprecise to be useful in monetary policymaking. Also, it would be appropriate to mention Laubach (2001) here because of some similarity of his empirical results to the papers mentioned in this paragraph. His Phillips curve relation takes the form:

$$\Delta \pi_t = \sum_{i=1}^{3} \beta_i \Delta \pi_{t-1} + \sum_{i=1}^{3} \gamma_i \left(u_{t-i} - u_{t-i}^N \right) + \sum_{i=1}^{2} \delta_{1,i} \Delta^2 s_{t-i}$$

$$+ \sum_{i=1}^{2} \delta_{2,i} \Delta^2 cp_{t-i} + \eta_t$$

where the natural rate of unemployment is assumed to take a random walk with or without a drift; a variable taking a random walk with a drift has a first time difference consisting of a stationary variable and (basically) a constant. Also, s means the nominal exchange rate, and cp commodity prices. His initial motivation is an interesting one: that he tries to derive the NAIRU for the G7 countries, except Japan, as well as for Australia. When a drift accompanies the random walk, it is assumed that the drift itself can follow a random walk. Although his estimation of the latter case (involving a random-walking drift) has improved the precision, the paper generally concludes that his NAIRU specification does not explain the joint evolution of unemployment and inflation.

One question that would have to be raised here is: although the Phillips-inflation equation in the above paragraph are carefully formulated, why does he use the change in the inflation rate as the explanatory variable? Is it not the rate of inflation which is sup-posed to respond to the gap between the actual unemployment rate and the NAIRU? Staiger-Stock-Watson's and Laubach's explanation is that expected inflation in period t is the actual inflation in period $t-1$, which gives rise to the Phillips relation with a first difference in inflation on the left-hand side. In other words, they assume a static expectation formation, which generally is not rational, as well as is too crude compared to other parts of the models. (Also, if their rea-soning is correct, Δs and Δcp should not be differenced again to get second differences of s and cp.)

The derivation of NAIRU in the work so far surveyed more or less leans toward using time series analysis and the definition of the NAIRU only, hence, as is correctly described by Laubach (2001, p. 230), it lacks much economic content. Rissman (1986) first decom-poses the unemployment rate into three components: the cyclical one, the frictional one, and the structural one. The last two parts compose the natural unemployment rate. Using the OLS regression she then estimates the cyclical unemployment rate as a function of current and past values of GDP, money supply, and the total number of the unemployed, among others. (GDP and money supply here mean the unexpected parts of the whole quantities.) Subtracting the cycli-cal component from the total unemployment rate gives the natural rate.

Attempts at estimation of potential output using Kalman filtering are not large in number. Kuttner (1992, 1994) are among the few. His model in 1992, rather than that in 1994, better conforms to our discussion that underlines the common aspects of the estimation of NAIRU and potential output. His 1992 model consists of the three equations. The first describes the growth of real GDP:

$$\Delta x_t = a_0 + a_1(x_t^* - x_t) + a_2 \Delta x_{t-1} + a_3 \Delta x_{t-2} + e_t^d + a_4 e_{t-1}^d$$

where x_t is the natural log of real GDP at time t, x_t^* is potential output, which is either the output level forthcoming with the normal use of labor and capital stock or the output level which brings the firm (sec-tor) the maximum profits at prevailing prices so that the firm wants to maintain that output level. Also, e_t^d is a period t shock to real GDP that does not affect potential output.

The second component concerns the changes in inflation:

$$\Delta \pi_t = b_1(x_{t-1} - x^*_{t-1}) + b_2(x_{t-2} - x^*_{t-2}) + b_3\Delta(m_{t-1} - x_{t-1} - p_{t-1})$$
$$+ e^\pi_t + b_4 e^\pi_{t-1},$$

where $m_{t-1} - x_t - p_{t-1}$ is the lagged logarithm of the inverse of $M2$ velocity, capturing the direct effect of money stock changes on inflation changes. Potential output is assumed to follow a random walk with a constant drift (g):

$$x^*_t = x_{t-1}{}^* + g + e^s_t.$$

He estimates the above system using US quarterly data from 1960:1 through 1991:3, and obtains significant values for important coefficients (as well as others), such as a_1, b_1, b_2, b_3, and g.

Although Clark (1987, 1989) did not exactly develop models for the NAIRU and potential output, it seems appropriate to refer to his models here for a reason that will shortly become clear. First, for output lyr_t (the log of real GDP), he decomposes it into a stochastic trend component n_t and a stationary cyclical component s_t. He then assumes n_t follows a random walk with a drift. The drift is also assumed to take a pure random walk (without a drift). He next assumes s_t to take a second-order autoregressive process. Second, for the unemployment U_t, he assumes it to consist of a trend component L_t and a stationary cyclical component C_t, where, it is assumed, L_t takes a pure random walk, while C_t follows a second-order autoregressive process. He then estimates these four unobserved components, n_t, s_t, U_t, and L_t, using the Kalman filtering technique. Since he calls the Okun Law the relationship between the gap ($lyr_t - n_t$: the cyclical component of output) and another gap ($U_t - L_t$: the cyclical component of unemployment), one might correctly guess he takes n_t and L_t as the NAIRU and potential output, respectively. He then obtains a main conclusion that the stationary cyclical components in both the product and labor markets are persistent and dominant in most of the industrial countries he examined. Decompositions of the movement of actual output into a trend and a cyclical component were exactly what Schumpeter (1935) was interested in. Clark's main finding reminds us of Schumpeter's observation that business cycles are the main dynamic activity in capitalist development, while economic growth is a concomitant and secondary phenomenon. It is also noted that Clark's conclusion is at variance with many other recent works, an example being Nelson and Plosser (1982), which emphasizes the (stochastic) trend component in most macroeconomic variables. Hence, similar work along those lines would be worthwhile.[3]

Let us close this brief survey with several remarks. The first is that, when one is concerned with time series regression to estimate the NAIRU or potential output, one is advised to make use of and incorporate the definition of NAIRU or potential output; for example, for the NAIRU, one has to formulate the inflation equation in such a way that when expectations on inflation (and on other variables) are fulfilled, the actual unemployment rate should revert to the NAIRU. The second point to note is that, since time series analysis and Kalman filtering can be used without going into details of the economics behind the estimation, it would be of interest and value to cross-check the results by using other methods, which focus more on economics. In this regard, Rissman's (1986) method may be usefully recalled; or, to put it another way, use of time series analysis and other statistical methods in a complementary way will result in a new and interesting perspective.

4.3 Japan's NAIRU and potential output, 1950–2002

4.3.1 The NAIRU and natural level of ROA (job offers/job applications)

In this section, we attempt to estimate Japan's postwar NAIRU and potential output using the Kalman filtering method. Three price indexes are used to express inflation rates; these are the GDP deflator, the wholesale price index, and the consumer price index (see Data Appendix for the sources of these indexes).

Let us first show two basic relationships for estimating the NAIRU and potential output. We write the observation equation involving the inflation rate and the excess demand for labor and so on, and next the two transition equations, as

$$p_t = a_1[nur_{t-1} - ur_{t-1}] + p_t^e + a_2(gimp_t - gimp_t^e) + u_t,$$

$$u_t \sim \text{iid } N(0, \sigma_u^2),$$

$$nur_t = g_{t-1} + nur_{t-1} + v_t, \ v_t \sim \text{iid } N(0, \sigma_v^2),$$

$$g_t = g_{t-1} + w_t, \quad w_t \sim \text{iid } N(0, \sigma_w^2),$$

$$(4.2)$$

where p_t is the rate of inflation in one of the price levels mentioned above, nur_{t-1} is the NAIRU or natural unemployment rate in year $t-1$, which is an unobservable state variable to be estimated by the above equation system, ur_{t-1} is the actual unemployment rate in year $t-1$, p_t^e is the expected inflation rate, which is the fitted value of p_t in a regression with explanatory variables p_{t-1}, p_{t-2}, p_{t-3}, and a constant. If the error term of this regression has mean zero and is not correlated with other right-hand variables, as is assumed in the standard OLS, then p_t^e is a rational expectation of p (see, for example, Lovell 1986). Also, $gimp_t$ is the inflation rate of the import price index (see the Data Appendix), while $gimp_t^e$ is the expected value of $gimp_t$, which is derived in the same way as for p_t^e. In the first equation of (4.2), inflation is assumed to respond to the excess demand of labor markets in the previous year. The error terms, u_t, v_t, and w_t, are assumed to be independent (of all other variables) and identically distributed, and have a normal distribution with zero mean and constant variances, σ_u^2, σ_v^2, and σ_w^2 (that is, they follow white noise processes). The second and third equations are the transition equations. The second equation shows nur_t which follows a random walk with drift g_{t-1}. The drift term is assumed either as also taking a random walk without a drift (the third equation), or as a non-zero constant or zero (in the last case, the third equation is not necessary).

The first equation says that when the actual inflation equals the expected inflation ($p_t = p_t^e$), and also actual imported inflation is equal to the expected rate (that is, unexpected inflation in the prices of imported goods is zero), then the actual unemployment rate equals the NAIRU (or natural unemployment rate). This state also implies that, since the labor market is in equilibrium with a 'natural slack' and expectations on inflations are fulfilled, the inflation rate is stable. This does not, however, imply that imported inflation is also stable, because this rate is not solely dependent on domestic market situations.

We will also estimate the natural level of the ratio of job offers to job applications (*roa: Yuko Kyujin Bairitsu* in Japanese) at public employment security offices (*Shokugyo Anteisho*). Here the excess demand index for labor in year t becomes $roa_t - nro_t$, where nro_t is the natural level of roa_t (or NAIRROA to use an analogy of NAIRU). It is now a well-known fact that in Japan, because of the long-term employment system (or the lifetime employment system), the unemployment rate may not properly reflect excess demand conditions in the labor markets. Hence, roa_t, which is composed of '*ex ante*' variables, has been used to represent the labor market tightness in Japan.

The second equation system below is for measuring potential output and is applied to one of the three inflation indexes. Representing a general inflation index by p_t as before, the second system consists of:

$$p_t = b_1(lyr_{t-1} - lpo_{t-1}) + p_t^e + b_2(gimp_t - gimp_t^e) + x_t,$$

$$x_t \sim \text{iid } N(0, \sigma_x^2),$$

$$lpo_t = h_{t-1} + lpo_{t-1} + y_t, \ y_t \sim \text{iid } N(0, \sigma_y^2), \tag{4.3}$$

$$h_t = h_{t-1} + z_t, \ z_t \sim \text{ iid } N(0, \sigma_z^2),$$

where, as in the previous case, h_t can be one of the following: a non-zero constant, zero, or a stochastic variable taking a random walk. The term lyr_t is the natural log of real GDP, lpo_t is potential output in natural logarithm, which is the state variable here, so that $lyr_t - lpo_t$ is the excess demand rate for commodities in year t. The second and third equations (transition equations) imply that potential output follows a random walk with a variable or a constant drift. A constant drift case means that z_t is identically zero with $h_0 \neq 0$.

The first maximum likelihood estimation (Model 1; the maximum likelihood method will be used throughout this and the following chapter), which is a variant of equation (4.2), yields

$$a_1 = 5.547 \ (7.123), \ a_2 = 0.129 \ (11.961),$$

$$\text{log-likelihood } (ll \text{ hereafter}) = 81.82,$$

where we assume that $\sigma_u^2 = 2 \times 10^{-4}$, $\sigma_v^2 = 10^{-5}$. The random walk in the natural unemployment rate is assumed to involve no drift term. Partly because of a limited number of sample points, we have to give variances of u_t and v_t the above exogenous values. The term σ_v^2 has the same value as the sample variance of the first difference of ur. We gave σ_u^2 a maximum number that yields a significant z value for a_1 (so that most smaller σ_u^2s than the above make a_1 significant).[4] The two numbers in the parentheses are z values (that is, the ratios of coefficients and standard errors when the standardized coefficients have a normal distribution); the corresponding p-values are both zero. A p-value is the probability of committing errors when one adopts the coefficient's values as true ones (that is, the probability of making the

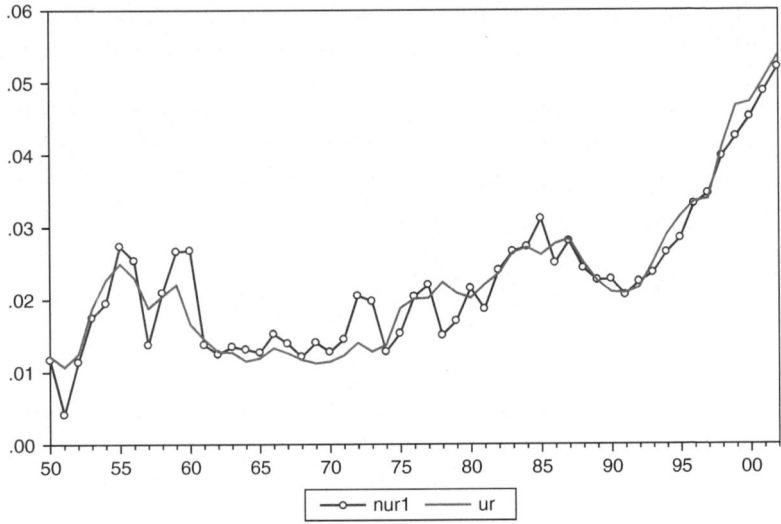

Figure 4.1 Natural and actual unemployment rates with the GDP deflator

type II errors). Figure 4.1 shows *nur*1 and *ur*, where *nur*1 denotes the natural rate of unemployment for Model 1. The excess demand for labor (*nur*1-*ur*) prevails in 1958–74, 1984–85, 1989–92, and briefly in 1997. Those years are mostly in the phases of business cycle expansions, the exception being 1992 which is on an early downturn of the 11th postwar inventory cycles. The peak of the 11th occurs in 1991 (see, for example, *Toyokeizai Economic Statistics Annuals*, Toyokeizai, Tokyo).

The second estimation (Model 2) uses *roa-nro*2 in the previous year as an excess demand variable for labor, where *nro*2 is Model 2's natural *roa* and is the state variable to be estimated. The variable *nro*2 is assumed to follow a random walk without a drift because the first difference of *roa* is stationary at the 1% level, according to the augmented Dickey-Fuller test. The result turned out as

$$a_1 = 0.0519 \ (8.133), \ a_2 = 0.105 \ (11.246), \quad ll = 95.072,$$

where we set $\sigma_u^2 = 2 \times 10^{-4}, \sigma_v^2 = 0.05$. The second variance was based on the sample variance of *roa*'s first difference, which is 0.049. For the first variance, we gave a consideration similar to Model 1's

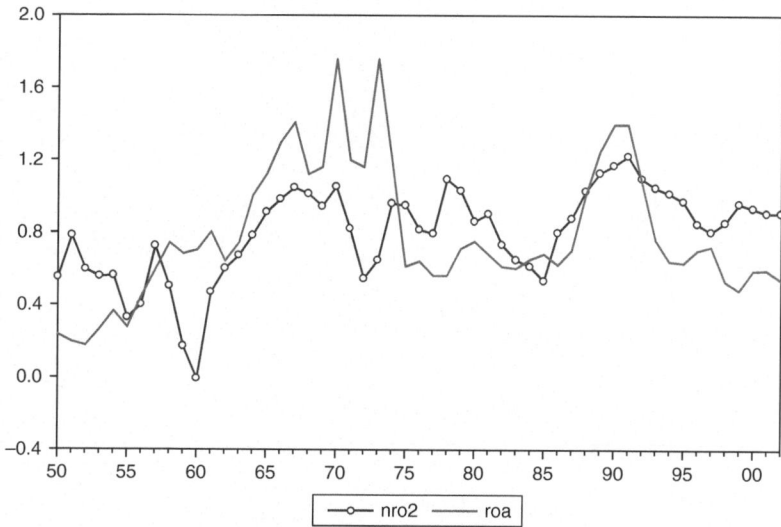

Figure 4.2 Natural and actual *roa*s with the GDP deflator

counterpart. The *roa* and estimated natural *roa*, *nro2*, are exhibited in Figure 4.2. The signs of excess demand for this case, *roa* − *nro2*, closely parallel those in Model 1; 1956, 1958–74, 1984–5, and 1989–91 have excess demand. In 1956 there was excess demand in Model 2, but there was not in Model 1. In Model 2 1992 and 1997 did not have excess demand, but in Model 1 it occurred in these years.

When one uses the rate of change of the domestic wholesale price index, *gwp*, as the measure of inflation, one observes that the excess demand for labor changes its sign much more frequently. Assuming $\sigma_u^2 = 2 \times 10^{-4}$ and $\sigma_v^2 = 4 \times 10^{-5}$, Model 3 is run to yield

$$a_1 = 5.102 \ (11.567), \quad a_2 = 0.365 \ (20.175), \quad ll = 82.565.$$

The natural rate of unemployment *nur3* and the actual rate *ur* are shown in Figure 4.3. The excess demand occurs in 1955–6, 1958–60, 1962, 1964–6, 1968–9, 1971–4, 1977, 1979–80, 1985, 1987–90, 1992, and 1997–8. Except for 1962, which is a bottom (first) year of the fifth inventory cycle, it is noted that all the other years are in expansionary phases or near the peaks of cycles.

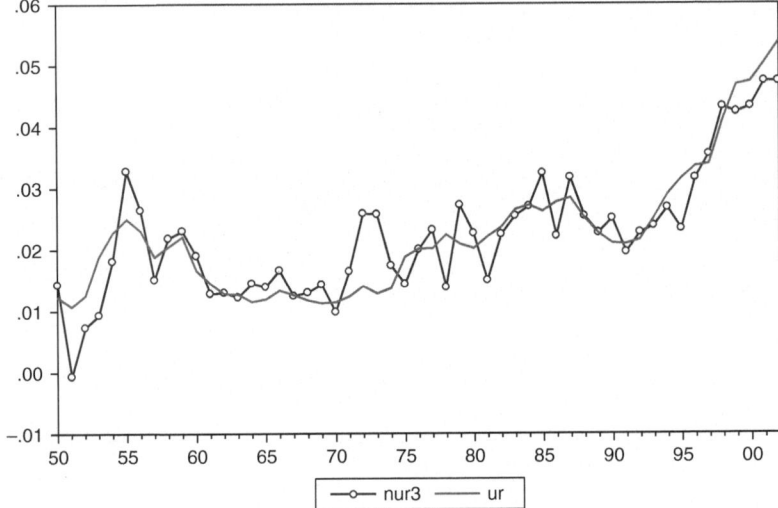

Figure 4.3   Natural and actual unemployment rates with the wholesale price index

As already noted, the excess demand in Model 3 changes its sign more often than in Models 1 and 2, and in cases when the consumer price index is used (see below), hence we will not show the model with *gwp* and excess demand expressed in terms of *roa*. However, comparing the models using three different price indexes has to be left for future examination.

Model 4, which uses the inflation rate in the consumer price index and where the excess demand is measured by *nur4 – ur* in the previous year, is estimated as

$$a_1 = 2.989 \ (7.756), \ a_2 = 0.136 \ (12.279), \quad ll = 102.138,$$

where the random walk in *nur* does not involve a drift term, and the variances are assumed as $\sigma_u^2 = 10^{-4}$ and $\sigma_v^2 = 5 \times 10^{-5}$. The time shapes of *nur4* and *ur* are given in Figure 4.4. One can note the range of variation in the gap between *nur4* and *ur*, that is, in the excess demand for labor, is wider here than in Model 1.

The final model belonging to the group (4.1), Model 5, also uses inflation in the consumer price index but the excess demand for labor is now measured by the gap between *roa* and its natural level, *nro5*, in

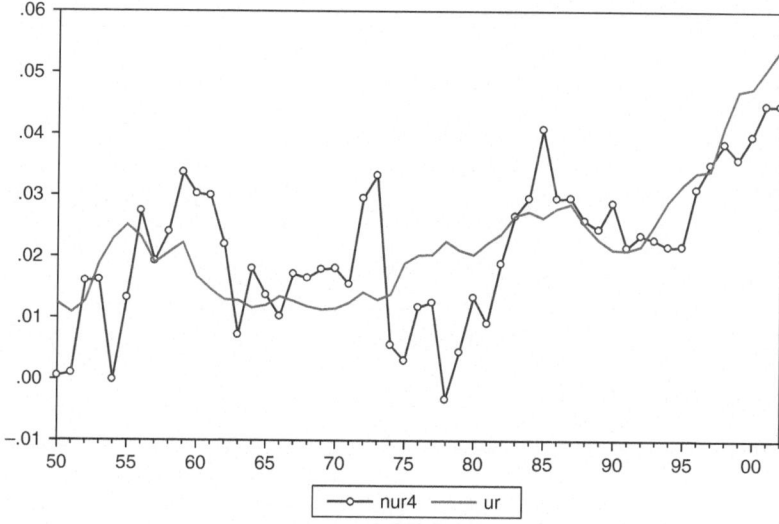

Figure 4.4 Natural and actual unemployment rates with the consumer price index

the previous year. The estimation was done to yield

$$a_1 = 0.017 \ (3.727), \ a_2 = 0.136 \ (10.664), \quad ll = 116.806,$$

where $nro5$ takes a random walk without a drift. We assume $\sigma_u^2 = 4 \times 10^{-4}$ and $\sigma_v^2 = 0.100$. Temporal changes in roa and $nro5$ are shown in Figure 4.5. Also, Figure 5a exhibits the time shapes of excess demand from Model 4 (EDL 4, where original numbers are multiplied by 92) and excess demand from Model 5 (EDL 5). Although $nro5$ takes a much smoother path than $nur4$, it can be seen that the shapes of excess demands, $nur4$-ur (EDL 4) and roa-$nro5$ (EDL 5) are quite similar. One can also observe that in the latter half of the 'high-growth period' (1965–73) and in the 1980s, EDL4 changes its direction more often than EDL5.

4.3.2 Potential output

Here we estimate potential output using several variants of equation system (4.2) and the three price indexes. The first is Model 6, where inflation is measured in terms of the GDP deflator and, in this case,

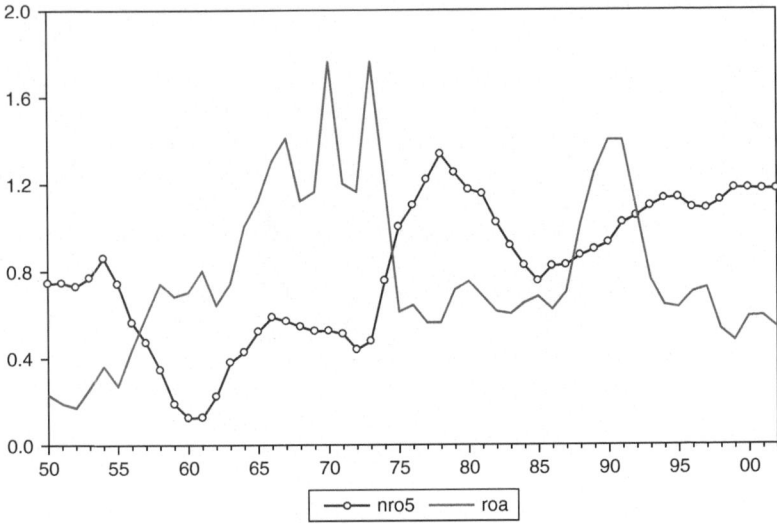

Figure 4.5 Natural and actual *roa*s with the consumer price index

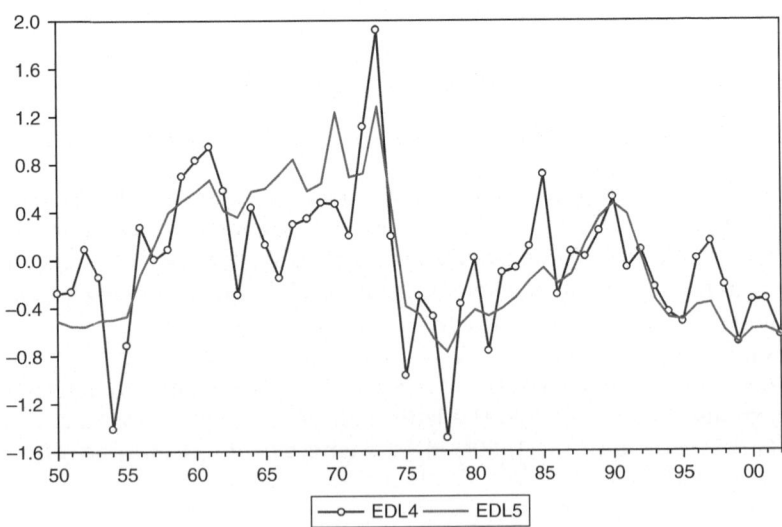

Figure 4.5a Excess demands from Model 4 and Model 5, with different measures of labor market gaps

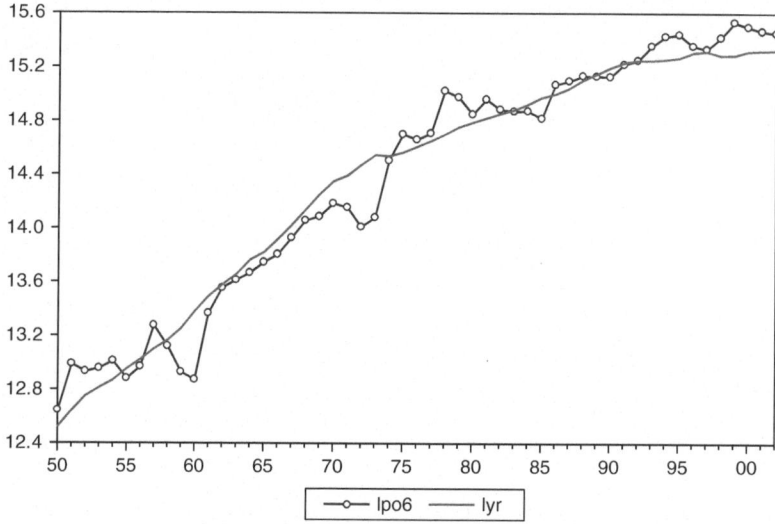

Figure 4.6 Potential and actual output with the GDP deflator

the excess demand is shown by the product market measure, *lyr* - *lpo6*. Model 6 is estimated as

$$b_1 = 0.097 \ (6.343), \ b_2 = 0.114 \ (8.960),$$

$$b_3 = 0.056 \ (1.720; \ p = 0.086), \quad ll = 98.158,$$

where b_3 is the estimated constant drift. We used a constant drift because the null hypothesis – that the first difference of *lyr* (approximately the growth rate of real GDP, *yr*) takes a random walk – cannot be rejected at the 5% significance level (it can be rejected at the 10% level). We assume $\sigma_x^2 = 2 \times 10^{-4}$ and $\sigma_y^2 = 0.05$. Log real GDP, *lyr*, and estimated potential output, *lpo6*, are shown in Figure 4.6. We also estimated the case where the drift follows a pure random walk, but the above estimation with a constant drift yields a better result in terms of the log-likelihood value.

Let us next use the domestic wholesale price index as a generator of inflation. In this case, Model 7, also assuming a constant drift rather than a random-walk drift, yielded a better log-likelihood. The coefficients and Figure 4.7 showing *lyr* and potential output *lpo7* are as

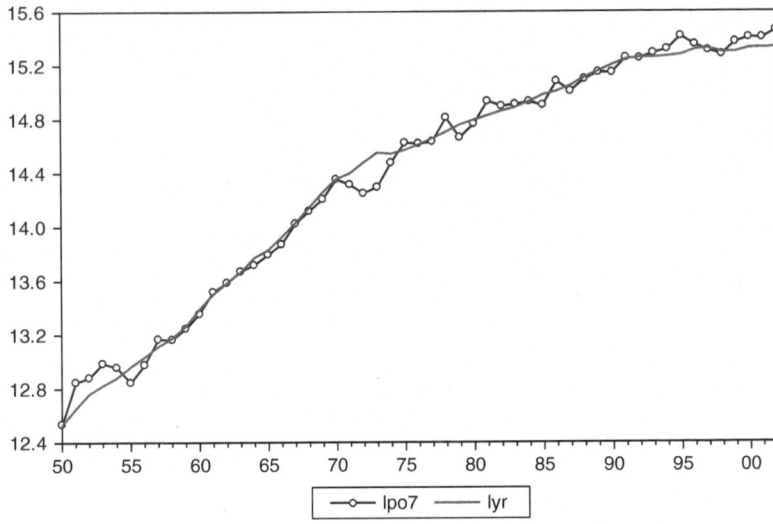

Figure 4.7 Potential and actual output with the wholesale price index

follows:

$$b_1 = 0.253 \; (7.827), \quad b_2 = 0.356 \; (22.555),$$

$$b_3 = 0.057 \; (4.406), \quad ll = 89.603,$$

where $\sigma_x^2 = 2 \times 10^{-4}$ and $\sigma_y^2 = 0.010$. The estimate of b_3 is for a constant drift term. Figure 4.7 implies that, as in Model 3, when the price level is the wholesale price index, the excess demand pressure during the high-growth period (1955–73) is weaker than in other models.

Finally, Model 8 uses the consumer price index as an inflation generator and, for the same reason as before, assumes a constant drift term rather than a random-walk drift. The results turned out as

$$b_1 = 0.136 \; (6.947), \quad b_2 = 0.133 \; (12.605),$$

$$b_3 = 0.055 \; (4.189), \quad ll = 108.130,$$

where $\sigma_x^2 = 2 \times 10^{-4}$ and $\sigma_y^2 = 0.010$. Real GDP, *lyr*, and potential output, *lpo8*, for this case (both in log), are shown in Figure 4.8. Also, Figure 8a displays the excess demand rate for commodities (EDC8) and excess demand for labor in Model 4 in terms of the rate of unemployment (EDL4, where the numbers of EDC8 for all years are multiplied by 23). The two diagrams are seen to evolve quite closely although the numbers of excess demand intervals are a bit different. One can observe that the EDC8 curve shows three intervals of years when excess demand for commodities prevails; these are 1956–74, which mostly conforms to the 'high-growth period'; 1984–5, which is the neighborhood of the peak of 10th business cycle; and 1987–92, which covers the upturn and one year after the peak of the 11th business cycle, whose peak occurred in 1991. The last interval mostly overlaps the 'bubble period' (1986–91).

4.4   Conclusions

This chapter consists of two parts. The first part was a survey of the literature on the theory and estimation of the NAIRU, or the natural rate of unemployment, and of potential output. These two concepts are important ingredients of monetary policy discussion because they are the levels compatible with stable inflation rates. Although the NAIRU was first proposed as a constant in a snap-shot image of the economy, when one views it as evolving over time it must change often according to changes in people's expectation, parameters defining the structure of the labor and commodity markets, the level of imported inflation, and so on. Most of the work surveyed uses the Kalman filter algorithm, which belongs to the group of state-space models or time-varying-parameter models, and which enables us to compute the above two magnitudes as functions of time. The chapter suggests that the future direction of work in this area would have to combine time series analysis and traditional economics as well as econometric considerations.

The second part of this chapter attempted to estimate Japan's NAIRU and potential output for 1950 through 2002. We proposed eight variants of the two basic equation systems, one using the gap between the NAIRU and the actual unemployment rate as the excess demand index, and other using the gap between the log real output and log potential output as the index. However, time patterns of excess demands from various models turned out quite similar, leading us to infer that our formulation of inflationary process mimics the actual evolution of the economy considerably well.

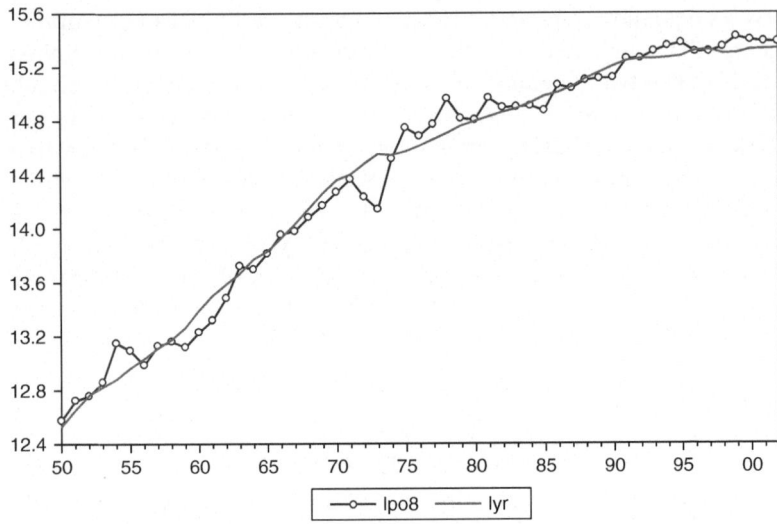

Figure 4.8   Potential and actual output with the consumer price index

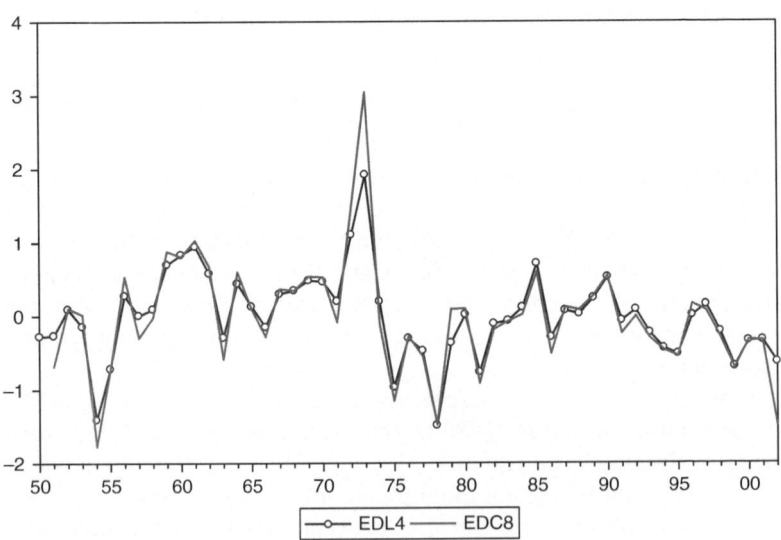

Figure 4.8a   Excess demands for labor and commodities from Model 4

The remaining tasks ahead of us would be (i) estimation of error variances in the observation equation and transition equations; (ii) refinement of the observation equation, particularly the appropriate embedding of shocks from abroad; (iii) extension of estimation to quarterly data and to cross-country comparison. We use here annual data to get round seasonal adjustment and choice of method involved therein. We would argue, however, that our estimation so far conducted makes a good starting point for further extensions and refinements.

List of Symbols

p	rate of inflation of the commodity price level.
nur	natural rate of unemployment.
NAIRU	non-accelerating inflation rate of unemployment, synonymously used with *nur*.
ur	actual rate of unemployment.
gimp	rate of inflation of the import price index.
roa	ratio of job offers to job applications at public employment security offices (in Japan).
nro	natural level of *roa*.
lyr	real GDP in natural logarithm.
po	potential output.
g, *h*	variables taking random walks with or without a drift, which either takes a pure random walk (without a drift) or is a constant.

Data Appendix

The GDP deflator is derived from nominal GDP/real GDP, both of which are derived from the *National Income Accounting Annual*, the Economic Planning Agency. The domestic wholesale price index draws on the *Monthly and Annual Statistics on Price Indexes*, the Bank of Japan. The import price index, the yen basis, is also derived from the same source. The consumer price index is for all of Japan and is comprehensive, and is derived from the *Consumer Price Index Annuals*, the Management and Coordination Agency.

The unemployment rate draws on the *Labor Force Survey Report* of the above agency. The ratio of job offers to applications is derived from the *Labor Statistics Annual* of the Ministry of Labor.

Notes

1. Fisher (1926, 1973) is sometimes mentioned as a predecessor having found the relationship earlier on. Amano (2003, ch 1) discusses a position of the Phillips curve in monetary policy discussion.
2. A random walk process has a first order difference which is stationary. A stationary time series has a constant mean and variance, and also its autocovariance depends only on the two time points of the variable.
3. Although Clark's work is an interesting and important one, we were a bit surprised to know that his business cycle analysis is conducted purely within time series methodology, without resorting to tools and notions in economics. One point of concern in his 1989 paper is that his diagrams showing Japan's cyclical components in output and employment for 1965–86 do not well mimic actual cyclical activities in this interval.
4. Inspection of graphs of excess demands in labor and commodity markets suggests that the first variance might be smaller than we assume here. But the size of the first variance does not seem to affect the points (years) when excess demands change their signs. In any case, estimation of σ_u^2 and σ_v^2 along with the coefficients would have to be our future task.

References

Amano, M. (2003) *Money, Inflation, and Output: A Study in International Perspective*. Chiba: Chiba University Research Monograph Series in Economics 5.

Bernanke, B.S. and F.S. Mishkin (1997) Inflation targeting: a new framework for monetary policy? *Journal of Monetary Economics* 11, 97–116.

Clark, P.K. (1987) The cyclical component of US economic activity. *Quarterly Journal of Economics* 102, 797–814.

——— (1989) Trend reversion in real output and unemployment. *Journal of Econometrics* 40, 15–32.

Fisher, I. (1926) A statistical relation between unemployment and price changes. *International Labor Review* 13, 785–92. Reprinted in *Journal of Political Economy* 81, 1973, 496–502.

Friedman, M. (1968) The role of monetary policy. *American Economic Review* 58, 1–17.

Gordon, R.A. (1977) A skeptical look at the 'natural rate' hypothesis. In: H.I. Greenfield et al., eds. *Theory of Economic Efficiency: Essays in Honor of Abba P. Lerner*, pp. 46–61. Cambridge, Mass.: MIT Press.

Gordon, R.J. (1997) The time-varying NAIRU and its implications for economic policy. *Journal of Economic Perspectives* 11, 11–32.

Hamilton, J.D. (1994) *Time Series Analysis*. Princeton: Princeton University Press.

Harvey, A.C. (1993) *Time Series Models*, 2nd ed. London: Harvester Wheatsheaf.

Kim, C.-J. and C.R. Nelson (1999) *State Space Models with Regime Switching*. Cambridge, Mass.: MIT Press.

King, R.G., J.H. Stock, and M.W. Watson (1995) Temporal instability of the unemployment-inflation relationship. *Economic Perspectives of the Federal Reserve Bank of Chicago* 19, 2–12.

Kuttner, K.N.(1992) Monetary policy with uncertain estimates of potential output. *Economic Perspectives of the Federal Reserve Bank of Chicago* 16, 2–15.

———— (1994) Estimating potential output as a latent variable. *Journal of Business and Economic Statistics* 12, 361–7.

Laubach, T. (2001) Measuring the NAIRU: evidence from seven economies. *Review of Economics and Statistics* 83, 218–31.

Lovell, M.C. (1986) Tests of the rational expectations hypothesis. *American Economic Review* 76, 110–24.

Nelson, C.R. and C.I. Plosser (1982) Trends and random walks in macroeconomic time Series: some evidence and implications. *Journal of Monetary Economics* 10, 139–62.

Phelps, E.S. (1967) Phillips curves, expectations of inflation, and optimal unemployment over time. *Economica* 34, 254–81.

Phillips, A.W. (1958) The relation between unemployment and the rate of change of money wages in the United Kingdom, 1861–1957. *Economica* 25, 283–99.

Rissman, E.R. (1986) What is the natural rate of unemployment? *Economic Perspectives of the Federal Reserve bank of Chicago* 10, 3–17.

Samuelson, P.A. and R.M. Solow (1960) Analytical aspects of anti-inflation policy. *American Economic Review* 50, 177–94.

Schumpeter, J.A. (1935) The analysis of economic change. *Review of Economic Statistics* 17, 2–10.

Staiger, D., J.H. Stock, and M.W. Watson (1997a) The NAIRU, unemployment, and monetary policy. *Journal of Economic Perspectives* 11, 33–49.

———— (1997b) How precise are estimates of the natural rate of unemployment? In: C.D. Romer and D.H. Romer (eds) *Reducing Inflation: Motivation and Strategy*, pp. 195–242. Chicago: University of Chicago Press.

5
The NAIRU, Potential Output, and Okun's Law: Postwar USA, UK, and Japan

5.1 Introduction

In the previous chapter we estimated the NAIRU (non-accelerating inflation rate of unemployment, which is synonymous with the natural rate of unemployment)[1] and potential output for postwar Japan using annual data. In that chapter, we left the following three points for future work. These are (i) to examine the relative merits of using, as a generator of inflation, either the GDP deflator, the consumer price index, or the wholesale price index; since potential output is a concept relating to GDP, a reasonable choice would be the first one, but one may need a formal criterion for comparing among the three indexes, (ii) to endogenously estimate error variances of an observation equation and a transition equation (or transition equations) of Kalman filter models,[2] and (iii) to handle the quarterly data, which will facilitate the work mentioned in point (ii).

The purpose of this chapter has two components: The first is to estimate the NAIRU and potential output in postwar USA, UK, and Japan. The second task is to estimate Okun's Law for the three countries. Okun's Law has several versions, each having a slightly different interpretation. The original version says that a 1-percentage-point decrease (from a natural unemployment rate) in the observed unemployment rate raises a 3- (or 3.2-) percentage-point increase in the relative excess of actual output over potential output. This relationship can be written as

$$(po - yr)/yr = 3(ur - 0.04),$$ (5.1)

or alternatively,

$$(yr - po)/po = -3(ur - 0.04) \tag{5.1'}$$

where *po* is potential output (real potential GDP), *yr* is actual output (real GDP), *ur* is the actual unemployment rate (expressed as a fraction). A fraction 0.04 is a natural rate of unemployment (a NAIRU) assumed in (5.1), which is, actually, not constant over time as well as cross-sectionally. Number 3 in (5.1) is a hypothetical level and what is called the 'Okun coefficient', which will also be variable depending on the time and economy in question. Equation (5.1) is the original form that Okun proposed, while (5.1') is a form put forward in Hall and Taylor (1997). In this chapter we use the second form because *po* would be a better base to measure the gap rate than actual *yr*. However, the difference between (5.1) and (5.1') is quite small because the left-hand side of (5.1) is nearly equal to $\ln(po/yr)$ while that of (5.1') is nearly equal to $\ln(yr/po)$ (particularly when *po* and *yr* are close to each other), and $\ln(po/yr) = -\ln(yr/po)$, where ln stands for the natural logarithm.

Our estimation here targets the NAIRU and potential output, hence combining the two estimates will give the Okun coefficient (assumed to be 3 above) in the three countries for the postwar period. Also, we use annual data because of a new method of giving appropriate exogenous error variances while referring to a variant of Okun's Law (which is given in annual terms) due to Blanchard (1997), who uses observable variables on both sides of the law. In other words, we use observable sample standard deviations in Blanchard's Okun relation, as criteria for deciding error variances in our Okun relations, which are made up of unobservable variables.

The next section (Section 2) presents a general framework for estimating the NAIRU and potential output. Then, the estimating procedures and comparative interpretations follow, where use is made of the GDP deflator as the price index. Next, we use the results regarding the NAIRU and potential output to derive Okun's Law for the three countries. Section 3 concludes, and is followed by an appendix that uses the consumer price index, and then a description of the data used.

5.2 The NAIRU, potential output, and Okun's Law with US, UK, and Japanese evidence

We first present the two basic relationships for deriving the NAIRU and potential output, which are the variants of 'expectation-augmented'

Phillips curves. The first one involves the NAIRU, nur_t, as a state (unobservable) variable and is written as

$$p_t = a(1)(nur_{t-1} - ur_{t-1}) + p_t^e$$
$$+ a(2)(gimp_t - gimp_t^e) + u_t, \quad u_t \sim \text{iid} \quad N(0, \sigma_u^2),$$
$$nur_t = g_{t-1} + nur_{t-1} + v_t, \quad v_t \sim \text{iid} \quad N(0, \sigma_v^2),$$
$$g_t = g_{t-1} + w_t, \quad w_t \sim \text{iid} \quad N(0, \sigma_w^2),$$

(5.2)

where p_t is the rate of inflation of either the GDP deflator or the consumer price index. Hence we examine the two systems corresponding to the above two price indexes. The term nur_{t-1} is the NAIRU in year $t-1$, ur_{t-1} is the actual (observable) rate of unemployment, so that $nur_{t-1} - ur_{t-1}$ expresses excess demand for aggregate labor force. Also, p_t^e is the expected rate of inflation, which is an estimated value of a regression of p_t on p_{t-1}, p_{t-2}, p_{t-3}, and a constant. The term p_t^e is a rationally expected inflation when the error term has a usual property (see Amano (2003); Maddala (2001)). The term $gimp_t$ is the rate of inflation of the import price index, and $gimp_t^e$ is the expected value of $gimp$, which is derived in the same way as p_t^e.

The second equation in (5.2) shows that nur_t takes a random walk with a drift g_{t-1} (see note 2 of the previous chapter for the definition). This equation allows the drift to take a pure random walk (the third equation). But the second equation includes the case of nur_t taking a pure random walk, where g_{t-1} is identically zero, as well as the case of a constant drift, where g_{t-1} takes on a non-zero constant.

The variable nur_{t-1} is called the state (unobserved) variable; ur_{t-1} is called the observed variable (data). Also, the first equation is called the observation (measurement) equation, while the second and third equations are called the transition equations. The term 'iid $N(\cdot)$' on the right-hand side implies that each error term is distributed as an independent and identical normal.

The second system, which represents another 'expectation-augmented' Phillips curve, is composed of

$$p_t = b(1)(lyr_{t-1} - lpo_{t-1}) + p_t^e$$
$$+ b(2)(gimp_t - gimp_t^e) + x_t, \quad x_t \sim \text{iid} \quad N(0, \sigma_x^2),$$
$$lpo_t = h_{t-1} + lpo_{t-1} + y_t, \quad y_t \sim \text{iid} \quad N(0, \sigma_y^2),$$
$$h_t = h_{t-1} + z_t, \quad z_t \sim \text{iid} \quad N(0, \sigma_z^2),$$

(5.3)

where lyr_{t-1} is real GDP in year $t-1$ in natural log, and lpo is potential output also in natural log. All the other corresponding variables have similar meanings to those in (5.2). Here the state variable is lpo_{t-1}, and $lyr_{t-1} - lpo_{t-1}$ represents excess demand for aggregate commodities.

Based on the basic frameworks described above, we now estimate the two state variables of the USA, the UK, and Japan using the maximum likelihood method, with annual data for 1951 through 2002. When we tried to estimate two or three error variances in (5.2) or (5.3), they were not obtained with significant z-values within our current model settings (z-values are the ratios of coefficients and standard errors when the standardized coefficients have a normal distribution). Hence we need to give them exogenously. Here, Blanchard's (1997) version of the law can conveniently be called on. It is written as

$$ur_t - ur_{t-1}(\equiv \Delta ur_t) = -a(gyr_t - c)$$

or

$$gyr_t - c = -(1/a)(ur_t - ur_{t-1}), \tag{5.4}$$

where $(1/a)$ is the Okun coefficient, gyr_t is the growth rate of real GDP, and a constant c is what Blanchard calls a normal growth rate of output required to sustain a constant unemployment rate over time.

In other words, Blanchard uses the growth rate of real GDP minus a constant c for the relative output gap (compare to (5.1')), and the first difference of the unemployment rate for the difference between the actual rate of unemployment and its natural counterpart. Then, as long as c can be seen as a constant, one has

$$sd \text{ (relative output gap in (5.1'))} = sd(gyr),$$

and

$$sd(ur - nur) = sd(\Delta ur),$$

where subscript t is omitted from now on unless it causes confusion, and sd means a sample standard deviation. Note in the above relation that the left-hand sides involve unobservable variables, while the right-hand sides only observable variables. Then the standard deviations on the right-hand sides will give useful clues to decide the error variances of equations (5.2) or (5.3), which involve the variables appearing on the above left-hand sides.

We shall develop our discussion first using the GDP deflator as a generator of inflation, and next, briefly, using the consumer price index as an inflation generator in an appendix.

5.2.1 The NAIRU and potential output using the GDP deflator

We first deal with the US economy. The augmented Dickey-Fuller statistic with the null that Δura (that is the annual difference in the US unemployment rate ura) has a unit root is -6.166, while the 1% significance level is -3.563, the 5% level is -2.919, and the 10% level is -2.597. Hence the hypothesis is rejected at the 1% level. Setting $\sigma_u^2 = 10^{-5}$, $\sigma_v^2 = 2 \times 10^{-5}$ so that $\sigma_v = 0.0045$, and assuming a constant drift $a_A(3)$, model (5.2) for the US economy is computed as

$$a_A(1) = 0.712 \, (37.887), \, a_A(2) = 0.054 \, (14.506), \, a_A(3) = 0.007 \, (2.182).$$

A letter a or A (b or B; j or J) attached to constants or variables means it pertains to the USA (the UK; Japan, respectively). The numbers in parentheses are z-values. Also, $sd(ura - nura) = 0.0105$ correctly hits $sd(\Delta ura) = 0.0105$, where $nura$ is the US's NAIRU (natural unemployment rate); recall the correspondence between our formulation of Okun's Law (5.1') and that of Blanchard (5.4). The US NAIRU and ura are shown in Figure 5.1.

We next examine the UK economy. The augmented Dickey-Fuller test statistic with the null that the annual difference in the unemployment rate urb has a unit root turned out as -4.843, so the null can be rejected in any significance level (the three critical values are the same as in the USA). Here, we first tried a constant drift term, using the second error standard deviation which has a comparable size to $sd(urb - nurb)$, where urb and $nurb$ are Britain's actual and natural unemployment rates, respectively. But then, significant coefficients were not obtained. Hence we next assumed a system (5.2) with a stochastic drift, for $\sigma_u^2 = 10^{-5}$, and $\sigma_v^2 = \sigma_w^2 = 10^{-4}$ (and $\sigma_v = 0.0100$). The coefficients are estimated as

$$a_B(1) = 2.877 \, (11.809), \, a_B(2) = -0.130 \, (-3.290),$$

and also $sd(urb - nurb) = 0.0105$, which seems close enough to $sd(\Delta urb) = 0.0098$. The evolutions of $nurb$ and urb are shown in Figure 5.2.

Turning to the Japanese Phillips relationship involving unemployment gap (5.1), we first conduct the augmented Dickey-Fuller test with

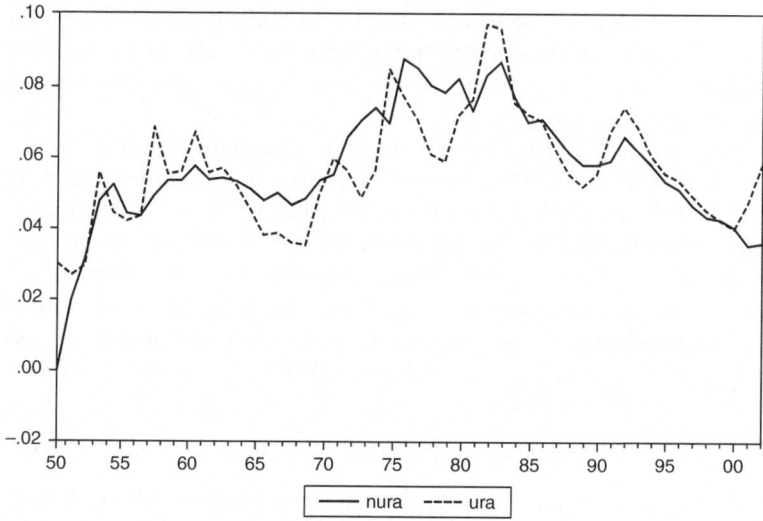

Figure 5.1 The US's natural and actual unemployment rates

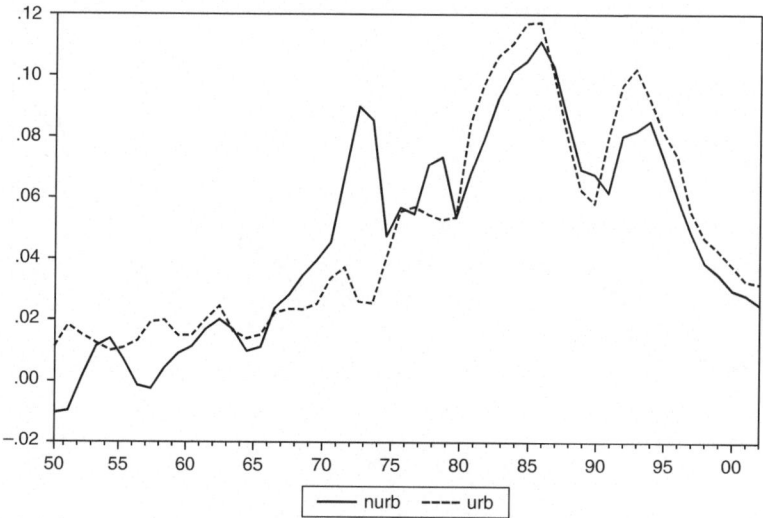

Figure 5.2 The UK's natural and actual unemployment rates

a null that Δurj has a unit root. The test statistic is −2.579; the three critical levels are the same as in the USA. Hence the null cannot be rejected at the 10% level (which, of course, does not necessarily mean $\Delta nurj$ has a unit root).

We tried in (5.2) the case where the first transition equation has a constant drift $a_J(3)$ and the case where it has a random-walk drift. In terms of the maximized log-likelihood, Akaike information criterion, and Schwartz criterion, the former turns out to have a better fit.

Regarding $sd(\Delta urj) = 0.0028$ as a reference value in choosing σ_v^2, while making $sd(urj\text{-}nurj)$ as close as possible to $sd(\Delta urj)$ (because of the correspondence between our formulation and Blanchard's), we set $\sigma_u^2 = 10^{-5}$ and $\sigma_v^2 = 10^{-5}$ ($\sigma_v = 0.0032$). Then, the maximum likelihood method yields

$$a_J(1) = 9.906\ (12.970),\ a_J(2) = 0.120\ (8.671),\ a_J(3) = 0.001\ (2.059),$$

and also, $sd(urj - nurj) = 0.0025$, which is reasonably close to $sd(\Delta urj)$, which is 0.0028. Note here that $nurj$ is an estimated series. The graphs of the estimated $nurj$ and actual urj are given in Figure 5.3.

We now turn to equations system (5.3), which expresses inflation as arising from the relative gap (excess demand) for commodities and

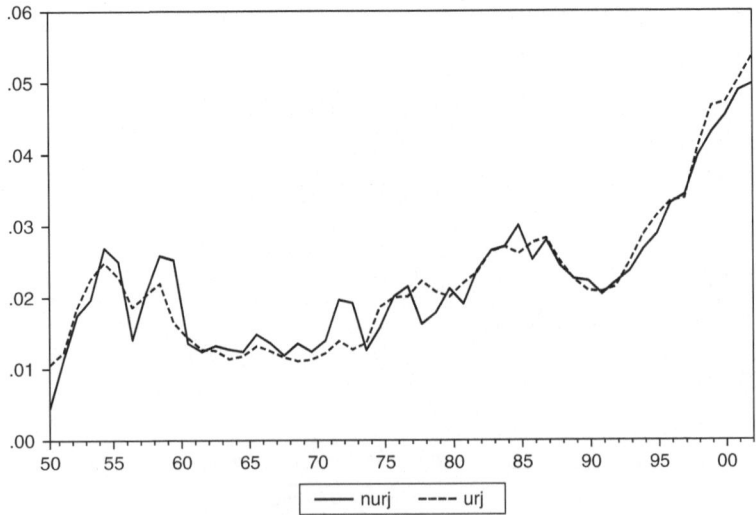

Figure 5.3 Japan's natural and actual unemployment rates

inflation expectation. For this case, we choose the two or three error variances such that the second standard error in (5.3) has a similar size to the standard deviation of output gap, and also that of output gap has a similar size to the standard deviation of GDP growth rates. The second criterion is due to the correspondence between our and Blanchard's Okun Law. The first one, however, is only a loose guidance, because the second error variance concerns differences in potential output, while the standard deviation of output gap involves the differences between actual output and potential output.

Starting with the US economy, we first obtain the augmented Dickey-Fuller statistic, which is -7.085, so that the null hypothesis for a unit root is rejected even at the 1% level. We assume a constant drift $b_A(3)$ for the first transition equation and choose $\sigma_x^2 = 1.5 \times 10^{-5}$, $\sigma_y^2 = 10^{-4}$, hence $\sigma_y = 0.0100$. Using those error variances, the system (5.3) yields

$$b_A(1) = 0.343\,(23.162),\ b_A(2) = 0.040\,(8.357),\ b_A(3) = 0.032\,(38.676).$$

Then, $sd(lyra - lpoa)$ is computed to be 0.0239, which is quite close to $sd(gyra) = 0.0235$, where $gyra$ is the growth rate of the US's output (real GDP). The first standard deviation implies that real GDP deviates from its potential by 2.4% on average. The evolutions of $lpoa$ and $lyra$ over the sample period are shown in Figure 5.4.

The derivation of real GDP from potential output for the UK proceeds in a similar manner. The test statistic of $gyrb$ for a unit root test turns out as -5.509, where $gyrb$ is the growth rate of Britain's GDP; hence the null hypothesis can be rejected at the 1% level. Referring to $sd(gyrb) = 0.0193$, we choose $\sigma_x^2 = 2 \times 10^{-5}$, $\sigma_y^2 = 3 \times 10^{-4}$ ($\sigma_y = 17.3 \times 10^{-4}$), with a constant drift term $b_B(3)$. The result turns out as

$$b_B(1) = 1.525\,(17.236),\ b_B(2) = -0.150\,(-6.395),$$

$$b_B(3) = 0.024\,(11.950).$$

The standard deviation of Britain's real growth rate $sd(gyrb)$ is 0.0193, while the standard deviation of her output gap $(lyrb - lpob)$ equals 0.0199, showing that they are close to each other. The latter also means that Britain's actual output deviates from her potential by 2% on average. The potential and actual output is drawn in Figure 5.5.

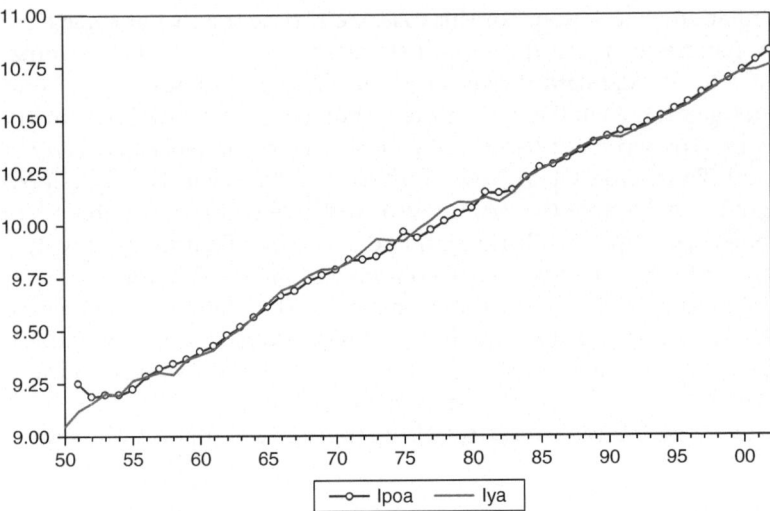

Figure 5.4   The US's potential and actual output

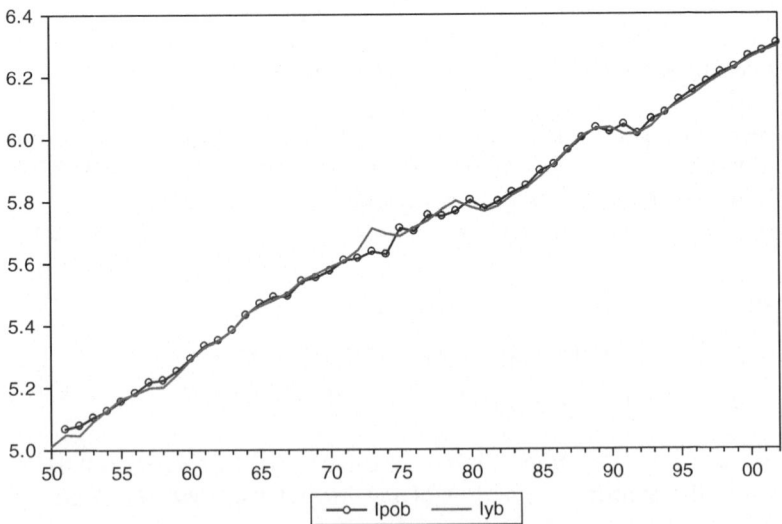

Figure 5.5   The UK's potential and actual output

Turning to the Japanese economy, the augmented Dickey-Fuller test statistic of $\Delta lyrj$ for the null hypothesis of a unit root is -2.511, so it cannot be rejected. We assume a constant drift term for the first transition equation, and set $\sigma_x^2 = 8 \times 10^{-5}$, $\sigma_y^2 = 1.5 \times 10^{-3}$ so that $\sigma_y = 38.7 \times 10^{-3}$.

Then the maximum-likelihood estimation using system (5.3) yields

$$b_J(1) = 0.611 \ (10.520), \ b_J(2) = 0.079 \ (5.429), \ b_J(3) = 0.052 \ (9.134).$$

Also, $sd(lyrj - lpoj) = 0.0414$, while $sd(gyrj) = 0.0402$, where $gyrj$ is the annual growth rate of real GDP, yrj. The two sds are considerably close to each other. Note also that $sd(lyrj - lpoj) = 0.0414$ implies that, on average, real output diverges from its potential level by 4.1%. The potential and actual output, $lpoj$ and $lyrj$, are drawn in Figure 5.6.

5.2.2 Okun's Law in terms of the GDP deflator

Let us now derive Okun's Law for the three countries. The US's Okun Law is derived by regressing $dyra \equiv lyra - lpoa$ on a constant and $dura \equiv ura - nura$, where the former (output gap) is the excess demand for output, while the latter (unemployment gap) is the excess supply

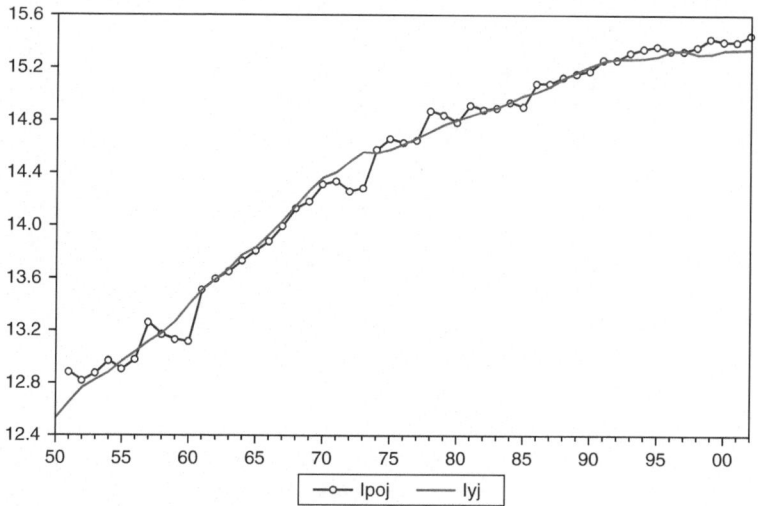

Figure 5.6 Japan's potential and actual output

of labor. In the following we omit writing estimated values of constant terms, although they are always present in the estimation. The US's Okun coefficient can be obtained as $-c_A(1)$ of the following OLS estimate:

$$dyra = c_A(0) + c_A(1)dura + e,$$

assuming, as in the other two countries, that the error term e satisfies the usual requirements.

The result turns out as

$$c_A(1) = -2.214 \ (31.797), \ \overline{r^2} = 0.952, \ DW = 2.260,$$

where $\overline{r^2}$ is the adjusted coefficient of determination for the degree of freedom, DW is the Durbin-Watson ratio, and the number in parentheses on the right-hand side is the t-ratio. Hence the US's Okun coefficient is 2.214. The relationship is drawn in Figure 5.7 below.

Secondly, the UK's counterpart is obtained from regressing $dyrb \equiv lyrb - lpob$ on a constant and $durb \equiv urb - nurb$. One then obtains

$$c_B(1) = -1.903 \ (-117.351), \ \overline{r^2} = 0.996, \ DW = 2.478.$$

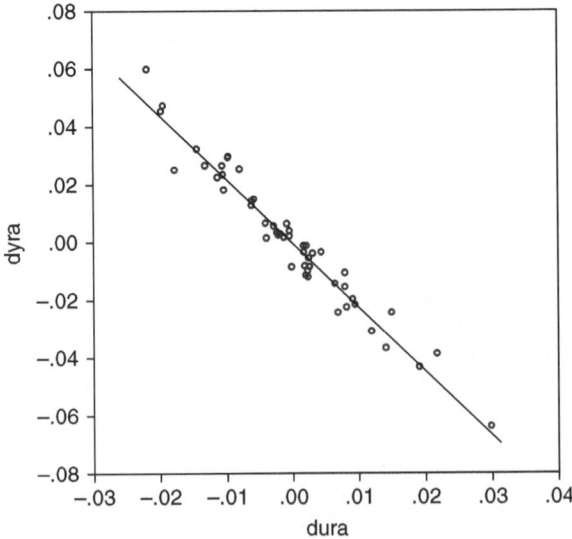

Figure 5.7 The US's Okun relationship

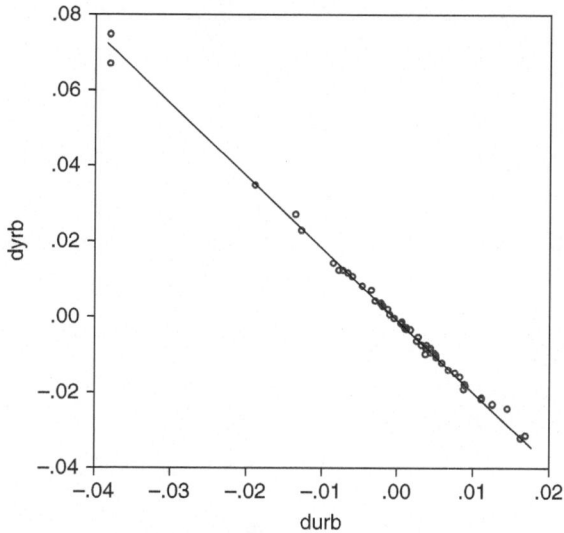

Figure 5.8   The UK's Okun relationship

The UK's Okun coefficient is therefore 1.903. See Figure 5.8 for the UK's Okun relationship.

Finally, we deal with the Japanese case. Writing $durj \equiv urj-nurj$ and $dyrj \equiv lyrj -lpoj$, the result for Japan is

$$c_J(1) = -14.276 \, (-26.239), \; \overline{r^2} = 0.931, \; DW = 1.709.$$

Hence the Okun coefficient is 14.276. The law is depicted in Figure 5.9 below.

It is of some interest to refer to the Okun coefficients that are found in other studies. Hamada and Kurosaka (1984) compute the Japanese coefficients for three subperiods between 1953 through 1982, which are 18, 32, and 13, the weighted average of which amounts to 21. They also derive the coefficient for the US for the same time period, which is 2.36. The big difference in the coefficients between Japan and other countries is, according to them, due to the different response of (average) labor productivity following a one-percentage decrease in unemployment. Blanchard (1997) extracts the coefficients for the US, the UK, and Japan, among others. As was described in (5.4) above, he uses output growth in excess of the normal growth rate c (which he

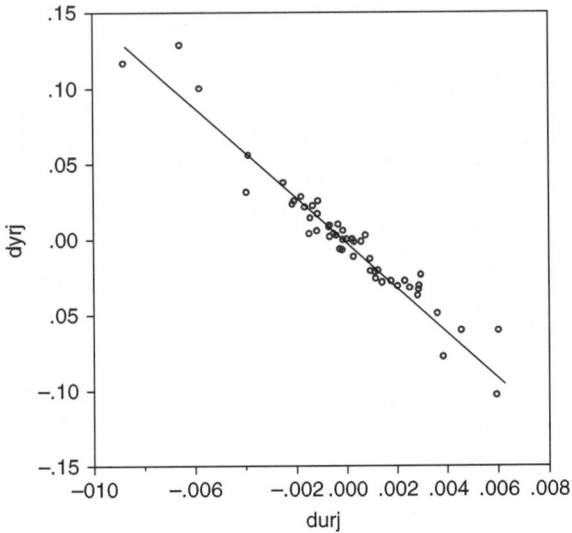

Figure 5.9 Japan's Okun relationship

regards to be 3%) in place of output gap, and the difference in the unemployment rate from the last year in place of the unemployment rate gap. His estimated coefficients are, on average over two subperiods for 1960 through 1994, 2.35 for the USA, 4.37 for the UK, and 5.75 for Japan. He attributes the larger coefficients in Japan and the UK to social or legal obstacles to firms' employment adjustments.

Comparing our estimates with those of the preceding authors, for the USA, the three estimates are close in size. The UK's coefficient in Blanchard (1997) is larger than twice the size of ours. For the Japanese economy, ours is located in between those of Hamada-Kurosaka and Blanchard. Note should be taken here that, in those studies, the estimation periods are different, some of which must involve drastic structural changes such as the collapse of the bubble in the Japanese economy in the early 1990s.

As is recognized by Hamada-Kurosaka and Blanchard, the Okun coefficient seems to be changing over time. Further inquiries into the size and international difference of the coefficient as well as its stability over time seem to be the subjects waiting for our future studies.

5.3 Conclusions

This chapter is an attempt to estimate the NAIRU and potential output for the postwar US, UK, and Japanese economies using the Kalman filter algorithm. The main device of our framework that enables the estimation results for the three countries to be comparable to each other is that we call on Blanchard's (1997) version of the Okun Law, where he contrasts output growth in excess of the normal growth rate and changes in the rate of unemployment. In the original Okun relationship, the two variables are the output gap in excess of its potential level and the rate of unemployment in excess of its natural level. Although this original version involves unobservable variables, Blanchard's version is entirely composed of observable variables. Using this correspondence we choose the two error variances in Kalman filter formulations to estimate the above two latent variables as well as Okun's relations for the three countries.

It will be in order and convenient to summarize our discussion and future directions of inquiry. The main task before us was the estimation of the NAIRU, potential output, and Okun's Law for the three countries, the USA, the UK, and Japan. The derivation of the Okun coefficient depends on the error variances one assigns to the observation equation and transition equation(s) in the equation systems for the Kalman filter. Hence we invoke Blanchard's (1997) version of the law, which comprises observable variables (the yearly increment of unemployment rate and yearly growth rate of output). We chose the two variances referring to (a) the standard deviations of the increment of unemployment rate and (b) the standard deviations of output growth, such that the standard deviation of our estimated unemployment rate gap becomes nearly equal to (a), and that of the estimated output gap is nearly equal to (b). Our devices are essentially for standardizing the three countries' results and for making the Okun coefficients comparable among them.

It will be plausible to assume the Okun coefficients to be variable over time, just like we started this study with the thought that the NAIRU (natural rate of unemployment) is not constant. Estimating the variable coefficient version might be well within our reach with the use of Kalman filtering.

It will also be our next necessary task to give economic explanations to the difference in the coefficients in various countries as well as to the levels and changes of the NAIRU. These aspects of future work seem much harder for us because they must involve legal and institutional (or even cultural) considerations of the product and labor markets in the economies under examination.

List of Symbols

yr	real GDP.
lyr	natural log of *yr*; the first *l* implies that the variable is in natural logarithm.
nur	natural rate of unemployment.
p	rate of inflation in the GDP deflator.
gimp	rate of inflation in the import price index.
po	potential output.
a, b, j	country attribution implying that the variable pertains to the USA, the UK, or Japan; for example, *poa* is the US's potential output.

Appendix. The NAIRU, Potential Output and Okun's Law for Consumer Price Indexes

The US Phillips curve involving the unemployment rate gap is derived using $\sigma_u^2 = 2.5 \times 10^{-5}$ and $\sigma_v^2 = 10 \times 10^{-5}$ ($\sigma_v = 32 \times 10^{-4}$). The resulting $sd(ura - nura) = 0.0106$, while $sd(\Delta ura) = 0.0105$. The US's other Phillips curve featuring the output gap has the error variances $\sigma_x^2 = 10 \times 10^{-5}$, $\sigma_y^2 = 55 \times 10^{-5}$ ($\sigma_y = 0.0235$). The estimated $sd(lyra - lpoa) = 0.0251$ while $sd(gyra) = 0.0235$. The last three magnitudes seem close enough to one another. The Okun coefficient can be given as $-c_A(1) = -1.898$ (-9.587), $\overline{r}^2 = 0.641$, $DW = 2.196$.

The UK's Phillips curve based on the unemployment gap, with a constant drift and plausible sizes of error variances, was not obtained with statistically significant coefficients. Hence we tried a random-walk drift in this Phillips curve; see equation (5.2). Here, we posit $\sigma_u^2 = 2 \times 10^{-5}$, $\sigma_v^2 = 1 \times 10^{-4}$, and $\sigma_w^2 = 6 \times 10^{-4}$ ($\sigma_y = 0.010$), while $sd(\Delta urb) = 0.0098$. Then the estimation gives $sd(urb - nurb) = 0.0141$, which is not close to $sd(\Delta urb) = 9.8 \times 10^{-3}$. However, the probability that $c_B(2)$ takes a wrong sign is 0.129. If one assigns a smaller value to σ_v^2 to make $sd(urb - nurb)$ closer to $sd(\Delta urb)$, then the above probability becomes larger, making $c_B(2)$ less significant. Hence we use the above error variances. For another Phillips curve using the output gap, we can use a constant drift. The error variances are $\sigma_x^2 = 7 \times 10^{-5}$, and $\sigma_y^2 = 3.5 \times 10^{-4}$ ($\sigma_y = 18.7 \times 10^{-3}$). The resulting standard deviation of output gap is $sd(lyb - pob) = 0.0193$, which exactly equals $sd(gyrb)$. The Okun coefficient then is given as

$$-c_B(1) = -1.361 \, (-57.200), \; \overline{r^2} = 0.985, \; DW = 2.468.$$

Our next target is Japan's Phillips curve involving unemployment gap that expresses excess demand pressures. Note, throughout this appendix, that the price level is measured by the consumer price index. Trying to make $sd(urj - nurj)$ and σ_v close to $sd(\Delta urj)$, which is 28×10^{-4}, we choose $\sigma_u^2 = 2 \times 10^{-5}$ and $\sigma_v^2 = 10^{-5}$. Then, $\sigma_v = 32 \times 10^{-4}$. Using these error variances along with a constant drift term, we run equation (5.2) and obtain $sd(urj - nurj) = 29 \times 10^{-4}$. The last standard deviation is reasonably close to $sd(\Delta urj)$ as well as to σ_v (of course, these two need not be very close because the former concerns the difference in actual unemployment rates, while the latter concerns the difference in NAIRUs).

All the computations in this appendix yield statistically significant coefficients at least at the 10% level except for Britain's $b_B(2)$.

Turning to equation (5.3), Japan's Phillips curve regarding the output gap is computed using two error variances $\sigma_x^2 = 2 \times 10^{-5}$ and $\sigma_y^2 = 2 \times 10^{-3}$. Here, $\sigma_y = 0.0447$. The standard deviation of the output gap is $sd(lyrj - lpoj) = 0.0438$, which is not very far from $sd(gyrj)$ ($= 0.0402$).

As in the text, the Okun coefficient can be extracted from the regression of the output gap on the unemployment rate gap and a constant. It turns out as

$$-c_J(1) = 14.646 \, (-33.098), \; \overline{r^2} = 0.955, \; DW = 1.825.$$

Comparing the two sets of Okun coefficients involving different price indexes, both the US and Japan have coefficients of considerably similar size, while the UK's coefficient using the consumer price index is much smaller than that using the GDP deflator. For the moment, it would be safe to pay attention mainly to the results in the text, because the consumer price index pertains only to a part of the aggregate commodity bundle, which comprises the whole gross domestic products.

Data Sources

All the data relating to the USA and the UK draw on *International Financial Statistics Yearbooks*, International Monetary Fund.

The data on Japan were derived from *The National Income Accounting Annuals*, Economic Planning Agency (for nominal and real GDP); *The Monthly and Annual Statistics on Price Indexes*, the Bank of Japan (for the domestic wholesale price index); *The Consumer Price Index Annuals*, the Management and Coordination Agency (for the consumer price index); *The Labor Force Survey Report*, the Management and Coordination Agency (for the unemployment rate).

Notes

1. The two terms are used in different contexts; the former is used in the context of anti-inflation policy while the latter mainly in a discussion of the shape of Phillips curves. See Johnson and Layard (1986) for an extensive discussion, particularly on the causes and international differences of the natural rate of unemployment.
2. See the previous chapter and Kim and Nelson (1999) for a comprehensive discussion and applications of Kalman filter models or, more generally, state-space models.

References

Amano, M. (2003) *Money, Inflation, and Output: A Study in International Perspective*. Chiba: Chiba University Research Monograph Series in Economics 5.

Blanchard, O. (1997) *Macroeconomics*. Upper Saddle River, N.J.: Prentice-Hall.

Hall, R.E. and J.B. Taylor (1997) *Macroeconomics*, 5th ed. New York: Norton.

Hamada, K. and Y. Kurosaka (1984) The relationship between production and unemployment in Japan: Okun's law in comparative perspective. *European Economic Review* 25, 71–94.

Johnson, G.E. and P.R.G. Layard (1986) The natural rate of unemployment: explanation and policy. In: O. Ashenfelter and P.R.G. Layard (eds). *Handbook of Labor Economics, Vol.II*, pp. 921–99. Amsterdam: Elsevier Science.

Kim, C.-J. and C.R. Nelson (1999) *State Space Models with Regime Switching*. Cambridge, Mass.: MIT Press.

Maddala, G.S. (2001) *Introduction to Econometrics*, 3rd ed. New York: Wiley.

Part III

6
Finance and Growth: VARs with Cointegration for the USA, the UK, and Japan

6.1 Introduction

A considerable literature has been developed so far, trying to find the effects of financial development on the real aspects of economic development and, particularly, on the growth of per capita real GDP, since the seminal work of Gurley and Shaw (1955, 1960). They argue that by offering more extensive sorts of saving instruments to households, the development of financial institutions (FIs) can increase saving flow from households and raise firms' investment volume, with broader menus of loan packages and with risk reduction through the economy of scale and risk-pooling. The development can also enhance the quality (profitability) of investment by specialized screening techniques of FIs. In summary, the development of FIs occurs in advance of, and assists, the development of real economic activity through those avenues.

The above temporal and causal relationship may be called the 'Gurley-Shaw hypothesis' because they are probably the first to bring economists' attentions to that causal pattern, although the naming is not a conventional one. More than two decades later, the same idea was put into econometric testing by Kormendi and Meguire (1985) and Barro (1991), which is now called cross country growth regression. All these regression analyses, as well as a recent one, namely Beck, Levine, and Roayza, (2000), detect temporal precedence (and causality) of financial development over real-side development (that is, per capita GDP growth). Barro (1997) and Levine (1997) offer a useful overview, and Levine and Renelt (1992), Quah (1993) and Mankiw (1995) present reviews on the literature from some critical angles.

99

Almost independently of this growth regression work, some papers in a different vein appeared, though smaller in number than growth regressions, using time series analysis and the concept of Granger causality (Granger 1969). They are concerned with causality relationships between financial development and real-side progress (GDP growth), where the measures of financial development take on various forms, typical ones being $m2$/GDP (Marshallian $k2$) and commercial banks' claims (or deposits)/GDP ratios, where $m2$ is broader money supply including time deposits. One of the earliest papers in this group is Jung (1986) who shows, using Granger causality tests (F-tests), that causality from financial development to real-side progress was observed in a majority of countries he dealt with. Similar efforts were made by Demetriades and Hussein (1996) and Neusser and Kugler (1998), where the former focuses on 16 developing countries while the latter is concerned with 13 OECD countries.

Later, starting from the Granger causality concept, Toda and Phillips (1993) develop a more efficient causality testing method than traditional Granger tests. Luintel and Kahn (1999), following Toda and Phillips' method, extend Granger tests to include cointegration and error-correction mechanisms and show that, for 10 developing countries, the causal direction is all uniformly bidirectional. Arestis et al. (2001) deal with five developed countries and point out that the directions of causality among developments of stock markets, banking, and the real-side (that is, per capita GDP growth) are different from country to country.

Arestis and Demetriades (1997) as well as Arestis et al. (2001) follow more closely the method of Johansen (1996). Christopoulos and Tsionas (2004) construct a panel-based vector error-correction and cointegration model and show that causality is unidirectional, running from financial development to economic growth.

In this chapter we develop a framework including cointegration and error-correction mechanisms that involve, to secure the identification of cointegrated relations, economically plausible cointegration coefficients (a few cases which exhaust all the plausible sets of coefficients), and estimate the above mechanisms simultaneously and, therefore, more efficiently (than previous two-stage estimation).

A new methodological device here is that, using economic and statistical arguments, it is made possible to examine all the (but at most a few) economically plausible sets of cointegrated relations and show that a representative cointegration set (whether the set includes two or three cointegrations) exhibits all the causalities that are included in other cointegration sets. As is shown formally by Wickens (1996) and Pesaran and Shin (2002), cointegrated relations cannot be identified without *a priori* restrictions on their coefficients, so that examining all the cases of economically plausible restrictions to be attempted here is a proper and relevant procedure. Another feature of time series analyses to be conducted below is that the measure of financial development includes direct finance, which is represented by the size of stock markets, in addition to indirect finance via banks.

Below, for prewar and postwar USA, UK, and Japan (which include six periods in total), using the above method, we examine the causal directions between financial development and real-side advancement. Also, referring to the results on causal directions, we examine whether the 'Patrick hypothesis' (Patrick 1966) is relevant to those countries. Although this naming is not an established one, it implies that in the early stage of development, financial development precedes and causes real-side development, while in the later stage, the causal direction becomes the reverse one. The hypothesis implying this reverse causality may be called the 'Robinson-Lucas hypothesis' because these authors, along with Kuznets, supported the latter causal order. In this regard, Robinson (1952) writes, 'by and large, it seems to be the case that where enterprise leads, finance follows,' and Lucas (1988) asserts that 'the importance of financial matters is very badly over-stressed'; see also Kuznets (1955).

The next section describes the building blocks, particularly just-identifying restrictions (on the coefficients of cointegrated relationships), which are representative of all the economically plausible cointegration sets, and examines whether the variables to be included in our analysis have a unit root. Section 3 explores the causal directions between financial and real development and asks if the Patrick hypothesis is relevant to the countries in question. Section 4 concludes with a summary and a few remarks.

6.2 Unit roots, cointegration, and error-correction

We consider VAR systems involving per capita real GDP (*lqp*), one of two measures representing the degree of financial development (*lsmp* and *lscp*; the definitions appear shortly), per capita capital formation (*lcp*), and per capita real exports (*lep*). We define *lsmp* as the natural log of [(the amount of annual stock transactions +*m*2)/population×(the GDP deflator)] for the USA and Japan, while for the UK the first term in the numerator is the values of annual stock issues. Stock transactions are also called stock market liquidity. The term *lscp* is the same as *lsmp* except for *m*2 being replaced by commercial banks' claims on private sectors.[1] Capital formation includes that of the government. In deriving real exports from nominal amounts, we used GDP deflators. The last two variables (*lcp* and *lep*) are included in the VAR system as 'exogenous demand factors'. The main symbols to be used in the text are described after Section 4. All the data are measured in annual frequency.

We examine the causal patterns between per capita income and financial development of the three countries for prewar and postwar periods. The prewar period for the USA is set from 1888 through 1940, the UK's counterpart is from 1888 through 1939, and that of Japan is from 1893 through 1940. The postwar periods run from 1946 through 1996 for the USA, from 1948 through 1999 for the UK, and from 1952 through 2003 for Japan. The reason for the difference in sample periods is mainly that the economic activity of the USA and the UK was disturbed much more severely by Great Depression in the early thirties, while the Japanese economy was thrown out of a normal track for several years following the end of the Second World War.

The maximum number of cointegrating vectors is derived in view of trace statistics, which are more robust than the method using the maximum eigenvalues (Cheung and Lai (1993)). There will be cases with one, two or three cointegrating vectors according to the criterion. Appendix A exhibits appropriate cointegration numbers at the 5% or 1% significance level for all the (six) cases. When the number of cointegrating vectors is at most two, and the financial variable is the sum of real per capita stock transactions and *m*2, we express the error-correction mechanisms as

$$
\begin{bmatrix} \Delta lqp_t \\ \Delta lsmp_t \\ \Delta lcp_t \\ \Delta lep_t \end{bmatrix} = \begin{bmatrix} \alpha_{11} & \alpha_{12} \\ \alpha_{21} & \alpha_{22} \\ \alpha_{31} & \alpha_{32} \\ \alpha_{41} & \alpha_{42} \end{bmatrix} \begin{bmatrix} 1 & 0 & -\beta_{13} & -\beta_{14} \\ 0 & 1 & -\beta_{23} & -\beta_{24} \end{bmatrix} \begin{bmatrix} lqp_{t-1} \\ lsmp_{t-1} \\ lcp_{t-1} \\ lep_{t-1} \end{bmatrix}
$$

$$(6.1)$$

where the first cointegrated relationship, whose matrix is in the middle of the right-hand side of (6.1), is

$$lqp_{t-1} = \beta_{13}\, lcp_{t-1} + \beta_{14}\, lep_{t-1}, \tag{6.2}$$

and the second one is

$$lsmp_{t-1} = \beta_{23}\, lcp_{t-1} + \beta_{24}\, lep_{t-1}. \tag{6.3}$$

In (6.1)–(6.3) a constant term is omitted, but when estimated it is present in all the equations. Also, a trend term may be added according to some criteria; Appendix B exhibits three criteria – the log-likelihood, the AIC, and the BIC – to decide whether a trend should be included in each cointegrated relationship.

Note that, in (6.1), the causality (in cointegrated relations) between per capita output and a financial variable is excluded to begin with; in that way, output and financial development are dealt with in an 'equal footing' manner. However, for causalities from all the other plausible cointegrating matrices see Appendix D, where it is shown that the cointegrated relations shown in (6.1) provide the most comprehensive causality patterns. The symbol Δ implies, as usual, a first difference.

The economic meaning of (6.2) is that there is a multiplier-type relation between capital formation and output production or, in terms of error-correction, the excess of per capita investment over per capita output prompts output growth; this mechanism is evidenced for many countries by DeLong and Summers (1991). Also, (6.2) implies that exports per capita will increase output per capita or, in error-correction terms, the excess of exports over output induces further output growth; see Levine and Renelt (1992). The meaning of (6.3) is that financial development will be spurred by higher per capita investment because this leads to the need for more finance. Besides, (6.3) implies that exports per capita, another exogenous demand element, may prompt financial development. In the case of two cointegrating vectors, we specify four βs in (6.1), where $4 = r^2$ and r is the number of cointegrating vectors as well as the rank of the cointegrating matrix (the middle matrix of Equation (6.1)). In this case, therefore, the system is *just-identified*. The two normalizing restrictions are put on *lqp* and *lsmp* (a financial variable) because of our focus on the causal patterns between output and a financial variable, and there needs to be only one long-run equation that explains each of per capita output and a financial variable.

According to Pesaran and Shin (2002, pp. 54–6), in order that the cointegrated relations be identified, r^2 restrictions must be distributed across the r different cointegrating vectors such that there are r restrictions for each of the r cointegrating vectors.

In the two cointegration case with four variables, each relation (row) must have $n - r = 2$ unrestricted variables, where $n\,(= 4)$ is the number of variables and $r(= 2)$ is the number of cointegrated relations (= the rank of the cointegration matrix), and $r^2\,(= 4)$ is the total number of restrictions to be distributed evenly to the two cointegrated relations (see Wickens 1996 and Pesaran and Shin 2002, who maintain that cointegrating relations cannot be identified by statistical arguments only, so that the identification requires *a priori* restrictions on the cointegration coefficients).

Now, to see that identified cointegrated relations are not unique, we show that the above two cointegration vectors can be derived from two general forms of cointegrated relations, using some manipulations that keep the linear independence of the two relations. The original general cointegrated relations can be written as

$$
\begin{bmatrix} 1 & -\beta_{12} & -\beta_{13} & -\beta_{14} \\ 1 & -\beta_{22} & -\beta_{23} & -\beta_{24} \end{bmatrix}
\begin{bmatrix} lqp_t \\ lsmp_t \\ lcp_t \\ lep_t \end{bmatrix} = \begin{bmatrix} 0 \\ 0 \end{bmatrix},
$$

where the two cointegrated relations are normalized on the first variable. (Normalizing the second row on the second variable does not change the following discussion.) Subtracting row 1 from row 2 of the above coefficient matrix yields

$$
\begin{bmatrix} 1 & -\beta_{12} & -\beta_{13} & -\beta_{14} \\ 0 & -\beta_{22}+\beta_{12} & -\beta_{23}+\beta_{13} & -\beta_{24}+\beta_{14} \end{bmatrix}.
$$

Next, we normalize row 2 by dividing the elements of row 2 by $-\beta_{22} + \beta_{12}$, which gives

$$
\begin{bmatrix} 1 & -\beta_{12} & -\beta_{13} & -\beta_{14} \\ 0 & 1 & -\beta_{23}{}^* & -\beta_{24}{}^* \end{bmatrix},
$$

where $-\beta_{23}{}^* = (\beta_{13} - \beta_{23})/(\beta_{12} - \beta_{22})$ and $-\beta_{24}{}^* = (\beta_{14} - \beta_{24})/(\beta_{12} - \beta_{22})$. Now, multiply the first column by β_{12} and add this to the second column to obtain

$$\begin{bmatrix} 1 & 0 & -\beta_{13} & -\beta_{14} \\ 0 & 1 & -\beta_{23}^* & -\beta_{24}^* \end{bmatrix}.$$

Then, renaming the βs in the second row, one obtains the middle term (cointegration matrix) of Equation (6.1). As the above derivation shows, however, one cannot restrict two cointegration vectors uniquely (that is, one can change the positions of zeros and βs without affecting the rank of the cointegration matrix). Hence, in Appendix D, we examine the causality results when the matrix takes all the other economically plausible forms (keeping output and financial development variables to be dealt with on an equal footing). Appendix D shows that the matrix used in the text yields the broadest causality patterns. See Enders (2010, ch 6) for a parallel argument.

The α_{ij}s are the speeds of adjustment (or loading factors) in cointegrated relations. For the first (second) cointegrated relation to be a meaningful one, one needs to have negative and significant $\alpha_{11}(\alpha_{22})$; see Luintel and Kahn (1999, pp. 388–9). A negative and significant α_{11} is neither a necessary nor a sufficient condition for the dynamic system to be stable (see, for example, Enders (2010, ch 6)). But a stable system with some significant $\alpha_{11}(> 0)$ will entail longer convergence time, which will render the steady-state with cointegrated relations non-practical to consider. Moreover, if $\alpha_{21} \neq 0$ and is significant, in the first cointegration, one can judge the causality runs from per capita GDP (the first variable in the VAR) to the financial variable (the second variable). When $\alpha_{21} = 0$, *lsmp* is said to be weakly exogenous with respect to *lqp*. If $\alpha_{22} < 0$ and is significant, and also if $\alpha_{12} \neq$, 0 and is significant, one can judge that causality exists, in the second cointegration, from the financial variable (the second variable) to per capita GDP (the first variable). When $\alpha_{12} = 0$, *lqp* is weakly exogenous with respect to *lsmp*. See Lutkepohl (2005, chs 6 and 7) for more discussion on the causality in VAR systems.

Toda and Phillips (1993) showed that joint estimation of cointegrated relationships (β_{ij}s) and error-correction mechanisms (α_{ij}s) as is done here can be carried out more efficiently than two-step procedures (that is, estimating β_{ij}s and α_{ij}s sequentially). They, as well as Hall and Milne (1994), Toda and Yamamoto (1995), and Luintel and Kahn (1999), develop discussions on causality detection and related topics in the context of generalized Granger causality tests, where 'generalized' means that the causality tests in VARs involve cointegration as well as error-correction mechanisms.

When the trace statistics imply that there are at most three cointegrating vectors, then the right-hand matrix of error-correction coefficients $[\alpha_{ij}]$ is 3×4, and the cointegrating vectors are specified here as

$$
\begin{bmatrix}
1 & 0 & -\beta_{13} & 0 \\
0 & 1 & -\beta_{23} & 0 \\
0 & 0 & 1 & -\beta_{34}
\end{bmatrix}. \tag{6.4}
$$

In (6.4), as in the two-cointegrated vector case, $r^2 = 9$ restrictions are distributed evenly across the three cointegrated vectors, with three constraints on each cointegrated vector.

Causality results emerging from two other plausible cointegrating vectors (these three exhaust all the plausible vectors; see Appendix E), where per capita output and financial variables are treated in an 'equal footing' manner, are described in Appendix E. In this appendix it is shown that the matrix in (6.4) yields the most comprehensive causality results.

As before, the three diagonal unities are normalizing restrictions, which are assigned to output per capita, a financial variable, and investment per capita, because one needs only one equation that determines each of the above variables in the long-run. The first (long-run) cointegrating vector in (6.4) implies the output-investment (multiplier) relation, the second means, as in the two cointegration case, that the financial development depends on per capita investment, and the third row indicates that the investment is prompted by higher export activity. Since the identifying restrictions are $r^2 = 3^2 = 9$, where r is the rank number as well as the number of cointegrated relationships, the cointegrating matrix (6.4) has *just-identifying* restrictions as in two-cointegration cases. It can easily be checked that if one reduces any one identifying restriction (that is, if one increases one β_{ij}), the system is unable to identify all the α_{ij}s and β_{ij}s simultaneously.

The following considerations, (i) through (iv), give the information to set up the most plausible identifying matrix for the three cointegration system: (i) the three main variables, lqp, $lsmp$ (or $lscp$), and lcp have a normalizing restriction, (ii) putting $\beta_{12} \neq 0$ and $\beta_{21} \neq 0$ cuts off the long-run interactions between the first two variables and the last two variables,[2] (iii) the effect of investment on output is stronger than that of exports on output (see Levine and Renelt (1992) for a supporting discussion), and (iv) it is plausible to posit that the effect on financial development of domestic capital formation is stronger than that of export activity.

In Appendix E, however, we examine the two alternative cointegration matrices (A1) and (A2), where (A1) violates supposition (iii), and (A2) contradicts supposition (iv); and we show that causalities derived from either (A1) or (A2) are the same, or included in, the causalities which result from the matrix in (6.4).

When the trace statistics indicate that the most probable number of cointegrations is at most three, the first two VAR equations with error-correction terms $(\alpha_{ij}[CRj], j = 1, 2, 3)$ can be written as

$$\Delta lqp_t = \alpha_1 + \Sigma a_{1k}\Delta lqp_{-k} + \Sigma a_{2k}\Delta lfd_{-k} + \Sigma a_{3k}\Delta lcp_{-k} + \Sigma a_{4k}\Delta lep_{-k}$$
$$+ \alpha_{11}(CR1) + \alpha_{12}(CR2) + \alpha_{13}(CR3), \qquad (6.5)$$

where $CR1 \equiv \gamma_1 + lqp_{t-1} - \beta_{13}lcp_{t-1}$, $CR2 \equiv \gamma_2 + lfd_{t-1} - \beta_{23}lcp_{t-1}$, $CR3 \equiv \gamma_3 + lcp_{t-1} - \beta_{34}lep_{t-1}$, and

$$\Delta lfd_t = \alpha_2 + \Sigma b_{1k}\Delta lqp_{-k} + \Sigma b_{2k}\Delta lfd_{-k} + \Sigma b_{3k}\Delta lcp_{-k} + \Sigma b_{4k}\Delta lep_{-k}$$
$$+ \alpha_{21}(CR1) + \alpha_{22}(CR2) + \alpha_{23}(CR3). \qquad (6.6)$$

(other two equations with Δlcp or Δlep as a dependent variable are omitted because they are not used in the following causality examination.) Here, k runs from 1 to an appropriate number which is indicated by the requirement that the VAR has some causality and that it passes the Lagrange multiplier test for serial correlation (see below where individual cases are examined; the AIC and BIC, which are used to seek optimal lag orders, tend to indicate optimal lags that are longer than those adopted here, but the VAR systems with longer lags generally give the same causalities as those shown here). lfd is one of the two financial development indicators, expressed in the logarithm. Also, α_i, α_{ij}, β_{ij}, a_{ij}, b_{ij}, and γ_i are all constants to be estimated. As Δ is a first difference, Δlfd, for example, stands for the growth rate of fd, where fd is either smp or scp.[3]

We next examine if the variables to appear in our VAR systems have a unit root (a stochastic trend) for each of the prewar and postwar periods. This test is important and conventional in finding cointegrated relations because it is necessary (but not sufficient) for there to be at least two $I(1)$ (integrated of order 1) variables if the variables have a cointegrated relation. An $I(1)$ time series can be made stationary if it is once time-differenced. The result of KPSS tests for unit roots (Kwiatkowski, Phillips, Schmit, and Shin (1992)) is shown in Appendix C, which is not attached to this chapter but is available from the author.[4] The tests reveal that for all the cases (two periods in each of the three countries), all cointegrated relations that are used for seeking causal directions have at least two $I(1)$ variables.

Also, note again that period divisions are different among the three countries because some exogenous (outside) events and war periods need to be excluded from the examination. We are now ready to examine the individual causality patterns in two periods for the three countries.

6.3   Causality tests on prewar and postwar US, UK, and Japanese evidence

In prewar USA, as Table 6.1 indicates, one cointegrated relation for financial variable *lsmp* was found, where, in the table, *lsmp* is shown after the period of analysis, indicating that the financial development led real-side advancement.[5] Hence the Gurley-Shaw hypothesis is relevant here. The sign and t-ratio for α_{11} indicate that the cointegrated relation is a meaningful one. Note that, because of the Lagrange multiplier test, the cointegrated relation has no serial correlation up to the indicated lags (in this case the significance level is 5%; see notes to Table 6.1). Also, in view of the White heteroskedasticity test, the residuals are homoskedastic. Hence the indicated relation does have white noise residuals, so that it can properly be called a cointegrated relation; all the cointegrated relations shown in the table turned out to have white noise residuals.

In the prewar US relationship, per capita capital formation and exports are estimated as substitutes, partly because of high correlations between these two (the correlation coefficient is 0.735). The opposite causality from output to either of the financial variables did not exist because only one cointegration in either case did not have a meaningful adjustment speed (that is, negative and significant α_{ii}).

Turning to postwar USA, we found one cointegration for financial variable *lsmp* and another one with *lsmp* on the left-hand side, which both have meaningful α_{ii} and are shown in the table.

Table 6.1 Cointegration, loading factors, and causalities

Prewar USA (1888–1940; *lsmp*, $l = 3$, $co = 1$, $plm = 0.090$, $pwh = 0.530$, a trend in the equation):

$$lqp = 0.184lsmp + 0.308lcp - 0.149lep + 0.010t,$$
$$\quad\;\;(9.500)\qquad\;(15.767)\qquad(6.299)\qquad(21.061)$$

$$\alpha_{11} = -1.590 \;(5.258).$$

Postwar USA (1946–96; *lsmp*, $l = 3$, $co = 1$, $plm = 0.518$, no trend):

$$lqp = 0.325lsmp + 0.252lcp + 0.121lep \;(pwh = 0.249),$$
$$\qquad(10.811)\qquad\;(3.465)\qquad\;(3.478)$$

$$\alpha_{11} = -0.356 \;(2.084).$$

$$lsmp = 3.081lqp - 0.775lcp - 0.374lep \;(pwh = 0.363),$$
$$\qquad(12.203)\qquad\;(3.108)\qquad\;(3.228)$$

$$\alpha_{11} = -0.343 \;(2.082).$$

Prewar UK (1888–1939; *lsmp*, $l = 6$, $co = 2$, $plm = 0.386$, $pwh = 0.492$, a trend in each equation):

$$\alpha_{11} = -0.370, \qquad \alpha_{21} = 0.777,$$
$$\quad\;(1.975)\qquad\qquad\;(1.737)$$

$$\alpha_{12} = -0.206, \qquad \alpha_{22} = -0.826.$$
$$\quad\;(1.405)\qquad\qquad\;(2.358)$$

Postwar UK (1948–99; *lscp*, $l = 4$, $co = 3$, $plm = 0.129$, 0.846, a trend in each equation):

$$\alpha_{11} = -0.487, \qquad \alpha_{21} = -4.549, \qquad \alpha_{31} = -0.423,$$
$$\quad\;(1.853)\qquad\qquad\;(3.494)\qquad\qquad\;(0.602)$$

$$\alpha_{12} = 0.016, \qquad \alpha_{22} = -0.255, \qquad \alpha_{32} = 0.026,$$
$$\quad\;(1.060)\qquad\qquad\;(3.411)\qquad\qquad\;(0.645)$$

$$\alpha_{13} = -0.072, \qquad \alpha_{23} = -0.500, \qquad \alpha_{23} = -0.204.$$
$$\quad\;(2.152)\qquad\qquad\;(3.015)\qquad\qquad\;(2.227)$$

(continued)

Table 6.1   (continued)

Prewar Japan (1893-1940; *lsmp*, $l = 6$, $co = 3$, $plm = 0.445$, $pwh = 0.254$, a trend in each equation):

$\alpha_{11} = -1.639$, $\alpha_{21} = -2.839$, $\alpha_{31} = -1.122$,
  (3.461)   (3.2281)   (0.669)

$\alpha_{12} = 0.258$, $\alpha_{22} = -1.517$, $\alpha_{32} = -0.212$,
  (1.994)   (6.406)   (0.463)

$\alpha_{13} = -0.177$, $\alpha_{23} = 0.382$, $\alpha_{33} = -0.108$.
  (2.171)   (2.570)   (0.374)

Postwar Japan (1952–2003; *lscp*, $l = 6$, $co = 3$, $plm = 0.201$, $pwh = 0.401$, a trend in each equation):

$\alpha_{11} = 0.820$, $\alpha_{21} = 6.361$, $\alpha_{31} = -0.542$,
  (2.449)   (3.320)   (0.660)

$\alpha_{12} = 0.088$, $\alpha_{22} = -0.268$, $\alpha_{32} = 0.561$,
  (4.195)   (2.131)   (2.770)

$\alpha_{13} = -0.816$, $\alpha_{23} = 4.816$, $\alpha_{33} = -0.753$.
  (3.029)   (3.124)   (1.139)

Notes: (1) *l*: lag order. (2) *co*: number of cointegrations. (3) *plm*: *p*-value in the Lagrange multiplier test for serial correlation, with a null hypothesis that there is no serial correlation in the residuals. (4) *pwh*: *p*-value in the White heteroskedasticity test (no cross terms) with a null of no heteroskedasticity. (5) Italicized coefficients and loading factors imply those are relevant to causality judgments. (6) Figures in parentheses are *t*-ratios in absolute value.

The latter cointegration has somewhat puzzling signs on the coefficients of per capita investment and exports. But per capita income has a correct and significant sign. Hence we use the second cointegration as a source of causality judgment.

In the UK prewar period, financial variable *lsmp* yields two cointegrated relations. The loading factors, shown in the table, indicate that the causality runs from per capita income to overall financial development *lsmp*. Here, two cointegrations have a meaningful α_{ii}, but, as their coefficients are not used in causality judgment, the table does not show the relations. Here the Robinson-Lucas hypothesis applies. The cumulative *t*-distribution at $F = 0.95$ for sample number 50 has a value 1.678, hence one can judge that α_{21}'s coefficient is significant, because the sign of the coefficient does not matter so that the test can be one-sided.

The postwar UK has three cointegrated relations, all of which are meaningful since α_{ii} ($i = 1, 2, 3$) are negative and significant at the 5% level. From the loading factors listed in the table, one can see that the causality runs from per capita output to financial variable *lscp*; that is the Robinson-Lucas hypothesis was relevant in postwar UK.

We now turn to prewar Japan. As is shown in the table, three cointegrations emerge, the first two of which are meaningful. In view of the loading factors, one sees that there was bidirectional causality between per capita income and overall financial development *lsmp*. Another financial development indicator *lscp* generated only one causal direction from output to financial development.

It remains to examine postwar Japan. For this period, three cointegrations were found, with the first two being not meaningful (α_{33} is negative but insignificant). The upshot is bidirectional causality between financial development *lscp* and output *lqp*; hence one cannot characterize postwar Japanese development as either conforming to Gurley-Shaw hypothesis or to Robinson-Lucas hypothesis. In this period, firms' insatiable demand for investment materialized only after being backed by bank loans, that is, firms were generally liquidity- (or investment funds-) constrained; simultaneously, this investment-led growth generated ample saving for its banking system to grow promptly.

The examinations so far conducted for the total of six periods are summarized in Table 6.2. What is apparent from the table

Table 6.2　A Summary of causality directions

Prewar	Postwar
USA: $l = 3$, $co = 1$, $lqp \leftarrow lsmp, lscp$ (GS)	$l = 3$, $co = 1$, $lqp \leftrightarrow lsmp$ (bidirectional)
UK: $l = 6$, $co = 2$, $lqp \rightarrow lsmp$ (RL)	$l = 4$, $co = 3$, $lqp \rightarrow lscp$ (RL)
Japan: $l = 6$, $co = 3$, $lqp \leftrightarrow lsmp$ (bidirectional)	$l = 6$, $co = 3$, $lqp \leftrightarrow lscp$ (bidirectional)

Notes: (1) *l*: lag order. (2) ←: causality from a financial development to per capita income. (3) →: causality from per capita income to a financial development. (4) ↔: bidirectional causality. (5) *co*: number of cointegrations. (6) GS: The Gurley-Shaw hypothesis is relevant. (7) RL: The Robinson-Lucas hypothesis is relevant.

is that causal directions in both prewar and postwar periods are quite country-specific and that the evolutions of causal patterns are also far from being similar among the three countries. Time series analyses of output-finance causality depend crucially on long-term data and statistical testing, and generally it is more often that this method does not share the results with cross-country regression analyses. Hence it might be a useful complementary work to time series analyses to follow and interpret what has been happening in each period of the three countries using knowledge of economic history.

Patrick (1968) suggested that in the early phase of economic development, finance tends to lead and cause real-side development (the Gurley-Shaw hypothesis), while in the later phase, the real-side tends to cause financial development (the Robinson-Lucas hypothesis). This reversal of causal ordering is called the 'Patrick hypothesis'. According to our analyses, the Patrick hypothesis does not seem to have a general validity. Only the US followed a pattern close to the hypothesis, but the other two countries exhibit quite different time shapes. Further examination will be needed, however, regarding the demarcation between the early and later periods as well as the extension of the number of sampled countries.

6.4　Conclusions

The purpose of this chapter has been to examine the causality relationships between financial development and per capita GDP growth for prewar and postwar USA, UK, and Japan, applying the newer method, which uses cointegration and error-correction mechanisms and which estimates the coefficients of both mechanisms at the same time and therefore efficiently.

A new methodological point of this chapter is that, applying the analyses of Wickens (1996) and Pesaran and Shin (2002) it was possible to formulate economically plausible and statistically significant sets of cointegrated (long-run) relationships and to choose a set that exhibits the most comprehensive causality results, both when the number of cointegrations is two and three. Another new feature of our time series analyses is that financial development includes direct finance, which is represented by stock market development.

The main observation from the above discussions is that causal directions between real development and financial progress are country- and period-specific, which in turn implies that the causal directions may depend on the specific historical paths each country followed, particularly those relating to the changes in institutional setting and economic policy. In view of the above outcomes, the next step would be to expand sample countries and compare the causality results to each other.

List of Symbols

lqp   per capita output in the natural logarithm; *l* means that it is a logged value.

lsmp   [stock transactions (or issues) + money supply *m2*]/(the price level × population).

lscp   the same as above except that *m2* is replaced by commercial bank claims.

lcp   real per capita capital formation.

lep   real per capita exports.

α_{ij} adjustment coefficient in error-correction mechanisms.

β_{ij} coefficient in cointegrated relations.

lfd   either *lsmp* or *lscp*.

t   time trend (1888 = 1).

Data Sources

The data used in this chapter were drawn from the following sources:

The USA

Gordon R.J., ed. (1986) *The American Business Cycle: Continuity and Change*. Chicago: University of Chicago Press.

International Monetary Fund (Various Years). *International Financial Statistics Yearbooks*. Washington, D.C.

US Department of Commerce (1975) *Historical Statistics of the United States*. Washington, D.C.

World Bank (2008) *World Development Indicators*. Online.

The UK

Capie, F. and Webber A. (1985) *A Monetary History of the United Kingdom, 1870–1982*. London: Allen and Unwin.

International Monetary Fund, op.cit.

Mitchell, B.R. (1988) *British Historical Statistics*. Cambridge: Cambridge University Press.

World Bank (2008) *World Development Indicators*. Online.

Japan

The Bank of Japan (Various Years) *Economic Statistics Annuals*. Tokyo.

Fujino, S. (1994) *Money Supply in Japan*. Tokyo: Keiso-shobo (in Japanese).

Hitotsubashi University Institute of Economic Research. *Long-Term Economic Statistics*, 1: K. Ohkawa et al. (1974) *National Income*; 2: M. Umemura et al. (1988) *Labor Force*. Tokyo: Toyokeizai-shinposha.

Management and Coordination Agency (1988) *The Historical Statistics of Japan*. Tokyo: Japan Statistical Association.

Notes

1. Stock transactions and stock issues are flow variables, while banks' claims and $m2$ are stock variables. Here, however, we interpret the latter two variables as service flows provided by these stocks, just as K in the production function represents a service flow delivered by a stock K, so that the sum of stock transactions (or stock issues) and $m2$ measure (direct and indirect) financial development. Similar measures of total financial development are employed in Beck, Demirguc-Kunt, Levine, and Maksimovic (2001, pp. 208–10).

2. In this case, the identifying matrix becomes $[1 \ -\beta_{12} \ 0 \ 0; \ -\beta_{21} 1 \ 0 \ 0; \ 0 \ 0 \ 1 \ -\beta_{34}]$, where, for example, the first four elements are the first row of a 4×3 matrix. Also, this matrix does not identify the VAR system, since causality between output per capita and financial development is reciprocal, and the matrix cannot lead to the unique estimation of coefficients.

3. When the cointegration number is two, $\alpha_{13} \equiv \alpha_{23} \equiv 0$ with $CR1$ and $CR2$ appropriately modified; see (6.2) and (6.3).

4. All the requests on omitted appendices and data used in this book should be addressed to masaamano70@gmail.com. Also, all the computations reported in this book were done with the econometric software package EViews, Ver. 6 (2007).

5. The cointegrated relation in Table 6.1 was found for lag order 3, but in the case of lag order 1 or 2, any meaningful cointegration, and hence any causality, was not found.

References

Arestis, P. and P.O. Demetriades (1997) Financial development and economic growth: assessing the evidence. *Economic Journal* 107, 783–99.

Arestis P., P.O. Demetriades, and K.B. Luintel (2001) Financial development and economic growth: the role of stock markets. *Journal of Money, Credit and Banking* 33, 16–41.

Barro, R.J. (1991) Economic growth in a cross-section of countries. *Quarterly Journal of Economics* 106, 407–43.

——(1997) *Determinants of Economic Growth: A Cross-Country Empirical Study.* Cambridge, Mass.: MIT Press.

Beck, T., A. Demirguc-Kunt, R. Levine, and V. Maksimovic (2001) Financial structure and economic development: firm, industry, and country evidence. In: A. Demirguc-Kunt and R. Levine, eds. *Finanial Structure and Economic Growth*, pp. 189–241. Cambridge, Mass.: MIT Press.

Beck, T., R. Levin, and N. Roayza (2000) Finance and the sources of growth. *Journal of Financial Economics* 58, 261–300.

Cheung, Y.-W. and K.S. Lai (1993) Finite-sample sizes of Johansen's likelihood ratio tests for cointegration. *Oxford Bulletin of Economics and Statistics* 55, 313–28.

Christopoulos, D.K. and E.G. Tsionas (2004) Financial development and economic growth: evidence from panel unit root and cointegration tests. *Journal of Development Economics* 73, 55–74.

DeLong, J.B. and L.H. Summers (1991) Equipment investment and economic growth. *Quarterly Journal of Economics* 106, 445–502.

Demetriades, P.O. and K.A. Hussein (1996) Does financial development cause economic growth? Time series evidence from 16 countries. *Journal of Development Economics* 51, 387–411.

Enders, W. (2010) *Applied Econometric Tme Series*, 3rd ed. New York: Wiley.

EViews, Ver. 6 (2007) Irvine, CA: Quantitative Micro Software,.

Granger, C.W.J. (1969) Investigating causal relations by econometric methods and cross-spectral methods. *Econometrica* 37, 424–38.

Gurley, G.S. and, E.S. Shaw (1955) Financial aspects of economic development. *American Economic Review* 45, 515–38.

——(1960) *Money in a Theory of Finance.* Washington, D.C.: The Brookings Institution.

Hall, S.G. and A. Milne (1994) The relevance of P-star analysis to UK monetary policy. *Economic Journal* 104, 597–604.

Johansen, S. (1996) *Likelihood-Based Inference in Cointegrated Vector Auto-Regressive Models*, 2nd ed. Oxford: Oxford University Press.

Jung, W.S. (1986) Financial development and economic growth: international evidence. *Economic Development and Cultural Change* 34, 333–46.

Kormendi, R.G. and P.G. Meguire (1985) Macroeconomic determinants of growth: cross-country evidence. *Journal of Monetary Economics* 16, 141–63.

Kuznets, S. (1955) Economic growth and income inequality. *American Economic Review* 45, 1–28.

Kwiatkowski, D, P.C.B. Phillips, P. Schmidt, and Y. Shin (1992) Testing the null hypothesis of stationarity against the alternative of a unit root. *Journal of Econometrics* 54, 159–78.

Levine, R. (1997) Financial development and economic growth: views and agenda. *Journal of Economic Literature* 35, 688–726.

Levine, R. and D. Renelt (1992) A sensitivity analysis of cross-country regressions. *American Economic Review* 82, 942–63.

Lucas, Jr., R.E. (1988) On the mechanics of economic development. *Journal of Monetary Economics* 22, 3–42.

Luintel, K.B. and M. Kahn (1999) A quantitative reassessment of the finance-growth nexus: evidence from a multivariate VAR. *Journal of Development Economics* 60, 381–405.

Lutkepohl, H. (2005) *New Introduction to Multiple Time Series Analysis.* New York: Springer.

Mankiw, N.G. (1995) The growth of nations. *Brookings Papers on Economic Activity.* No.1, 275–326.

Mayer, C. (1990) Financial systems, corporate finance, and economic development. In: R.G. Hubbard, ed. *Asymmetric Information, Corporate Finance, and Investment,* pp. 307–332. Chicago: University of Chicago Press.

Neusser, K. and M. Kugler (1998) Manufacturing growth and financial development: evidence from OECD countries. *Review of Economics and Statistics* 80, 638–46.

Patrick, H. (1966) Financial development and economic growth in underdeveloped countries. *Economic Development and Cultural Change* 14, 174–89.

Pesaran, M.H. and Y. Shin (2002) Long-run structural modeling. *Econometric Reviews* 21, 49–87.

Quah, D. (1993) Empirical cross-section dynamics in economic growth. *European Economic Review* 37, 426–34.

Robinson, J. (1952) The generalization of the general theory. In: J. Robinson, *The Rate of Interest and Other Essays.* London: Macmillan.

Toda, H.Y. and P.C.B. Phillips (1993) Vector autoregression and causality. *Econometrica* 61, 1367–93.

Toda, H.Y. and T. Yamamoto (1995) Statistical inference in vector autoregressions with possibly integrated processes. *Journal of Econometrics* 66, 225–50.

Wickens, M.R. (1996) Interpreting cointegrating vectors and common stochastic trends. *Journal of Econometrics* 74, 255–71.

Appendix A. Cointegration Rank Tests: Trace Statistics and Their 5% Significance Levels

		Trace statistics	5% critical values
Prewar USA:	$r = 0^*$	74.929	63.876
(1888–1940)	$r \leq 1$	24.186	42.915
Postwar USA:	$r = 0^*$	58.253	47.856
(1946–96)	$r \leq 1$	21.633	29.797
Prewar UK:	$r = 0^*$	90.089	63.876
(1888–1939)	$r \leq 1^*$	48.061	42.915
	$r \leq 2$	16.943	25.872
Postwar UK:	$r = 0^*$	86.945	63.876
(1948–99)	$r \leq 1^*$	49.073	42.915
	$r \leq 2^*$	26.562	25.872
	$r \leq 3$	8.791	12.518
Prewar Japan:	$r = 0^*$	163.427	63.876
(1893–1940)	$r \leq 1^*$	83.194	42.915
	$r \leq 2^*$	26.425	25.872
	$r \leq 3$	10.678	12.518
Postwar Japan:	$r = 0^*$	102.251	63.876
(1952–2003)	$r \leq 1^*$	44.784	42.915
	$r \leq 2$	22.562	25.872

Notes: The null hypothesis: Rank (the number of cointegration)= r.
*: Rejection of H_0 at the 5% level.

Appendix B. The Three Criteria for Setting Trends in Cointegrated Relations

	LL		AIC		BIC	
	No Trend	Trend	No Trend	Trend	No Trend	Trend
Prewar US	20.461	*238.164*	−6.055	*−6.685*	−3.825	*−4.418*
($co = 1$)						
Postwar US	382.380	*383.173*	−13.433	−13.424	−11.094	−11.046
($co = 1$)						
Prewar UK	418.806	*423.610*	−14.419	*−14.553*	−9.620	*−9.671*
($co = 2$)						
Postwar UK	327.447	*333.269*	−11.772	*−11.913*	−7.888	*−7.902*
($co = 3$)						
Prewar Jpn	316.978	*336.608*	−9.4114	*−10.225*	−4.231	*−4.917*
($co = 3$)						
Postwar Jpn	378.130	*380.874*	−12.92	*−12.953*	−10.271	−10.224
($co = 3$)						

Notes: (1) *co*: number of cointegrated relations. (2) The italicized figures indicate the chosen settings, which get at least two larger figures in absolute value in the three criteria.

Appendix C. Kwiatkowski-Phillips-Schmit-Shin (KPSS) Tests for Unit Roots

is omitted.

Appendix D. Alternative Cointegrated Relations When Their Number is Two

Here, we only state that when the cointegrating matrix for $co = 2$, which occurs only in postwar UK, takes either of these other two forms, the matrices do not yield any causality between output and financial development.

The other two matrices are:

$$\begin{bmatrix} 1 & -\beta_{12} & -\beta_{13} & 0 \\ -\beta_{21} & 1 & 0 & -\beta_{24} \end{bmatrix} \text{ and } \begin{bmatrix} 1 & -\beta_{12} & 0 & -\beta_{14} \\ -\beta_{21} & 1 & -\beta_{23} & 0 \end{bmatrix}.$$

Note that in these matrices, output and financial variables are treated in an equal footing manner ($\beta_{12} \neq 0$ and $\beta_{21} \neq 0$ in general; here, both terms are supposed to affect each other, if those cointegrated relations are estimated significantly).

Appendix E. Alternative Cointegrated Relations When the Number is Three

Here, we show that when $co = 3$, which occurs in three periods (see Table 6.2), the other two, less plausible cointegrating matrices yield causalities, which are the same as, or included in, the results from the matrix used in the text. These matrices are: $[1\ 0\ 0\ -\beta_{14};\ 0\ 1\ -\beta_{23}\ 0;\ 0\ 0\ 1\ -\beta_{34}]$ (A1) and $[1\ 0\ -\beta_{13}\ 0;\ 0\ 1\ 0\ -\beta_{24};\ 0\ 0\ 1\ -\beta_{34}]$ (A2), where, for example, the first four elements before the left-hand colon are those in the first row in the 3×4 matrices.

For matrix (A1), postwar UK shows causality $lqp \rightarrow lscp$, with $l = 4$, which is the same as in the text. In prewar Japan, one has $lqp \leftarrow lscp$, with $l = 6$, which is included in the bidirectional causality in the text. Also, postwar Japan provides bidirectional causality $lqp \leftrightarrow lscp$, with $l = 6$, which is the same as in the text.

In the case of (A2), postwar UK yields $lqp \rightarrow lscp$ causality, with $l = 4$, which is the same as in the text. Prewar Japan generates, for $l = 6$, $lqp \leftarrow lsmp$, which is included in the causality produced from the matrix in the text. Finally, postwar Japan posts bidirectional causality $lqp \leftrightarrow lscp$, with $l = 6$, which is the same as in the text and (A1).

7
Financial Structure and Economic Growth: Evidence from the USA, the UK, and Japan

7.1 Introduction

The past few decades have seen a considerable body of literature dealing with causality relationships between economic growth and financial development for a variety of countries, both developed and developing. In addition to the early classic work of Gurley and Shaw (1960), Goldsmith (1969), and McKinnon (1973), the econometric work of Kormendi and Meguire (1985) and Barro (1991) brought to light the important roles of financial development in raising growth rates of the economies. King and Levine (1993a, b) face more directly the role of financial development in growth processes. More recent work along this line, in addition, underlines the direct financial routes along with indirect ones in explaining the influence of financial development on economic growth; see Levine and Zervos (1998) and Rajan and Zingale (1998). As is well and usefully summarized in Demirguc-Kunt and Levine (2001, particularly chs 1, 3, and 5), this group of work showed in cross-country regression frameworks that it is mainly the total volume of finance and not the proportion of direct/indirect financial routes that is important for the subsequent growth performance of the countries. Beck and Levine (2002), using a panel data as well as cross-country regression, derive a conclusion similar to that of the above book that financial development, but not financial structure, positively affects the growth of industry and the economy.

Along with the above cross-country regression work, some time series analysts used the Granger causality test to examine if the causality extends from financial development to economic growth, or from the latter to the former. The mechanics behind the first causal direction may be called the 'Gurley-Shaw hypothesis' (see the first paragraph), and those behind the second the 'Robinson-Lucas hypothesis', because the latter authors thought that the causality from growth to finance is more normal than the other way round; see Robinson (1952) and Lucas (1988). The time series work that examines the causal direction between financial and physical development includes Demetriades and Hussein (1996), Neusser and Kugler (1998), and Luintel and Kahn (1999). The first two papers have shown that causal directions are country-specific, while the third, which examined 10 developing countries, exhibits that bidirectional causality prevails in all the countries they sampled. A recent paper of Luintel et al. (2008) samples 14 developing countries and uses time series and heterogeneous panel methods (particularly, fully modified OLSs) to show that financial structure as well as financial size affects the speed of GDP expansion of those countries.

This chapter develops and estimates four-variable VAR systems with cointegrated relations and error-correction mechanisms. The frequency of data is annual. The four variables we will deal with are per capita real GDP, the indicator of total size of direct and indirect financial development, the ratio between direct finance and indirect finance (financial structure), and real per capita capital formation.

A new device in the methodology of this chapter consists, as in the previous chapter, of applying some formal results proposed by Wickens (1996) and Pesaran and Shin (2002), which imply that for the identification of cointegration coefficients, one needs *a priori* economic theory-based restrictions on the coefficients. As is now well recognized, the (vector of) cointegration coefficients cannot be determined uniquely through statistical methods only; hence, we examine a few (exhaustive) economically plausible sets of coefficients, and then adopt a representative set that generates the most comprehensive causality results.

The purposes of this chapter are then to see the causal directions between economic growth and financial development and also whether financial structure, (the amount of direct finance)/(that of indirect finance), mattered for the growth rates of the economies we deal with. The subjects of this study are prewar and postwar USA, UK, and Japan, consisting of six periods in total. The method we will use yields efficient estimates of both coefficients of cointegration and error-correction by employing the framework developed by Toda and Phillips (1993) and Hall and Milne (1994). It estimates the long-run and short-run coefficients simultaneously (hence it is more efficient than previous ones estimating the two groups of coefficients in two stages), so that it may be called an extended (or generalized) Granger causality test, to be contrasted to traditional Granger tests. This chapter is a new attempt to examine the influence of financial structure on economic growth using a framework of extended Granger causality tests that involve cointegration and error-correction mechanisms. The broad message of this chapter is not at variance with the work of Luintel et al. (2008), but the analytical devices are different and the method used here may be better linked to the previous time series analyses, so that this chapter will be able to claim its own position among those dealing with similar topics.

Section 2 describes the VAR system we will handle for the above purposes. Section 3 applies the VAR systems to prewar and postwar USA, UK, and Japan to discuss causal directions between output (growth) and overall financial development, as well as whether changes in financial structure affect output (growth) for the two periods in the above three countries. Also, the causal patterns and directions in the sample countries and periods are compared with each other. Section 4 summarizes and concludes the chapter.

7.2   Unit roots, cointegration, and causality

The VAR system to be developed below has four variables: The first is per capita real GDP (*lqp*; the first *l* means it is in logarithm; this is also the case for other variables starting with *l*). The second variable is one of the two measures of financial development (*lscp*, *lsmp*, the descriptions of which appear shortly). The third variable is the ratio of direct finance to indirect finance, *lsc* or *lsm*; see below for their meanings. The fourth variable is per capita real capital formation, *lcp*, which measures an 'exogenous' demand factor, other than financial factors, that may affect output growth. Capital formation *lcp* includes that of the government. In the second (financial development) variables, *s* means the sales (or issues) of stocks in a year, *c* means bank credits to private sectors, and *m* is *m*2 money supply.[1]

Small letter s represents the direct financial route, while c and m represent indirect financial routes. Then $s+c$ and $s+m$ mean the measures of total financial development, which are divided by the GDP deflator and population size[2]; and finally, after taking logarithms, one obtains $lscp$ and $lsmp$; for example $lscp$ is a short-hand for $\ln[(s+c)/(\text{the GNP deflator} \times \text{population})]$.

In the third variable, sc or sm now means not the sum of $s+c$, and so on, but sc implies the ratio of stock sales (or stock issues) to bank credits, and sm the ratio of stock sales (or stock issues) to money supply $m2$.[3] These two ratios in logarithms, lsc and lsm, are intended to represent the financial structure, that is the ratio of direct (market) finance to indirect (bank) finance. The main symbols to be used in this chapter are gathered and explained in the List of Symbols following Section 4. Also, the data sources are described after the List of Symbols section.

The total periods of the three countries are divided into prewar and postwar subperiods. Prewar USA extends from 1888 through 1940, and its postwar period is from 1946 through 1996. The prewar UK period runs from 1888 through 1939, and its postwar period is from 1948 to 1999. The Japanese prewar period is from 1893 through 1940, while its postwar counterpart runs from 1954 through 2003. The difference in period divisions is due partly to data availability, but also, for the three countries, World War II and the reconstruction periods are excluded from the examination.

The frequently used criteria for deciding the number of cointegrating relationships are those using trace statistics and maximum eigenvalues, but the former is generally regarded more robust (Cheung and Lai (1993)). Appendix A to this chapter exhibits, for each period of the three countries, the trace statistics, their 5% critical values, and the maximum number of cointegrations which is significant at the 5% level.[4]

According to Pesaran and Shin (2002, pp. 54–6), in order that the cointegrated relations can be identified, r^2 restrictions must be distributed across the r different cointegrated vectors such that there are r restrictions (including one normalizing restriction) for each of the r cointegrated vectors. In the two cointegration case with four variables, each cointegrated vector (row) has $n - r = 2$ unrestricted variables, where $n(=4)$ is the number of variables, $r(=2)$ is the number of cointegrated vectors (= the rank of the cointegration matrix), and r^2 (= 4) is the total number of restrictions to be distributed evenly to the cointegrated relations (see Wickens (1996); Pesaran and Shin (2002, pp. 54–6 and 77)). They maintain that cointegrated relations cannot be identified (cannot be decided uniquely) by statistical arguments only, so that the identification requires *a priori* restrictions on the cointegration coefficients.

When the trace statistics indicate that the number of cointegrations is at most two, we write the error-correction mechanisms and cointegrated relations, as an economically plausible one, as

$$
\begin{bmatrix} \Delta lqp_t \\ \Delta lscp_t \\ \Delta lsc_t \\ \Delta lcp \end{bmatrix} = \begin{bmatrix} \alpha_{11} & \alpha_{12} \\ \alpha_{21} & \alpha_{22} \\ \alpha_{31} & \alpha_{32} \\ \alpha_{41} & \alpha_{42} \end{bmatrix} \begin{bmatrix} 1 & -\beta_{12} & -\beta_{13} & 0 \\ 0 & 1 & -\beta_{23} & -\beta_{24} \end{bmatrix} \begin{bmatrix} lqp_{t-1} \\ lscp_{t-1} \\ lsc_{t-1} \\ lcp_{t-1} \end{bmatrix}, \quad (7.1)
$$

where the first cointegrated relation is

$$
lqp_{t-1} = \beta_{12} lscp_{t-1} + \beta_{13} lsc_{t-1}, \tag{7.2}
$$

and the second relation is

$$
lscp_{t-1} = \beta_{23} lsc_{t-1} + \beta_{24} lcp_{t-1}. \tag{7.3}
$$

In the above two relations, a constant and a trend term is omitted, but when these are estimated, a constant is always present, and a trend is included in some relations; see below. With Δ being the difference operator as usual, the left-hand side of (7.1) means the growth rates of four non-logged variables.

Now, we can show that the above two cointegration vectors can be derived from two general forms of cointegrated relations, using some manipulations that keep the linear independence of the two relations. But as this point has been checked in Chapter 6, Section 2, readers who want to confirm it are referred to the previous chapter.

The economic meaning of (7.2) is that per capita output may be raised by total financial development per capita as well as by changes in financial structure. Note that, as for the effect on output, financial development and structure are treated in an 'equal footing' manner (that is, both are entered as a possible causal factor for output change). The second relation (7.3) implies that total financial development is spurred by some changes in financial structure and by more capital investment per capita, because the latter needs some form of extra finance. Since the number (r) of cointegrated relations is now two, the number of just-identifying restrictions should be $r^2 = 4$, among which two are the normalizing restrictions.[5]

For the cointegrating matrix (the middle matrix on the right-hand side of (7.1)) to be identified, β_{23} can be zero, which is called the over-identifying restriction. To obtain a small generalization here, we set $\beta_{23} \neq 0$.[6] The α_{ij} in (7.1) are the loading factors (or the adjustment coefficients), and the α_{i1} ($i = 1, \cdots, 4$) are the adjustment coefficients of the ith variable to the equilibrium error in the first cointegration, where, for example, $i = 1$ corresponds to lqp. The α_{i2} ($i = 1, \cdots, 4$) are defined similarly. If α_{11} is negative and significant, the first cointegration is meaningful; see Luintel and Kahn (1999). Additionally, if $\alpha_{21} \neq 0$ and is significant, one can judge that there is a causality from lqp (the 1st variable) to $lscp$ (the 2nd variable). When $\alpha_{21} = 0$, $lscp$ is weakly exogenous with respect to lqp; see Lutkepohl (2005, chs 6 and 7) for more discussion. Similarly, if α_{22} is negative and significant, then the second cointegration is meaningful, and also if $\alpha_{12} \neq 0$ and is significant, one obtains a causal pattern extending from $lscp$ to lqp. When $\alpha_{12} = 0$, lqp is weakly exogenous with respect to $lscp$.

When the trace statistics indicate that the maximum number of cointegrated relations is three, we set out the cointegrating matrix as follows:

$$\begin{bmatrix} 1 & -\beta_{12} & 0 & 0 \\ 0 & 1 & -\beta_{23} & 0 \\ 0 & 0 & 1 & -\beta_{34} \end{bmatrix}. \tag{7.4}$$

In (7.4), as in the two cointegration case, $r^2 = 9$ restrictions are distributed evenly across the three cointegrated vectors, with three restrictions (one being a normalizing restriction) on each cointegrated vector.

Omitting time subscripts, the first, second, and third cointegrated relations then read, respectively,

$$lqp = \beta_{12} lfd, \tag{7.5}$$

and

$$lfd = \beta_{23} lfs, \tag{7.6}$$

where *lfd* is one of the two logged measures of financial development, and *lfs* is one of the two logged measures of financial structure. Of course, the indirect finance measure, *c* or *m*, should be common on both sides of (7.6). Also,

$$lfs = \beta_{34} lcp. \tag{7.7}$$

Equation (7.5) implies that, in the long-run, per capita output is affected by financial development but not by financial structure. Equation (7.6) means that changes in financial structure prompt the overall financial development, and (7.7) means that increasing investment can be a source of changes in financial structure.[7] From (7.5) and (7.6), therefore, financial structure affects per capita output via overall financial development. In matrix (7.4), if some row has fewer restrictions (that is, more βs), one cannot identify all the αs and βs simultaneously.

When the trace statistics indicate that the number of cointegrations is at most three, the first two VAR equations with error-correction terms ($\alpha_{ij}[CRj]$, $j = 1, 2, 3$) can be written as

$$\Delta lqp_t = a_1 + \Sigma a_{1k}\Delta lqp_{-k} + \Sigma a_{2k}\Delta lfd_{-k} + \Sigma a_{3k}\Delta lfs_{-k} + \Sigma a_{4k}\Delta lcp_{-k}$$
$$+ \alpha_{11}(CR1) + \alpha_{12}(CR2) + \alpha_{13}(CR3), \tag{7.8}$$

where $CR1 \equiv \delta_1 + lqp_{t-1} - \beta_{12}lcp_{t-1} + \gamma_1 t$, $CR2 \equiv \delta_2 + lfd_{t-1} - \beta_{23}lfs_{t-1} + \gamma_2 t$, $CR3 \equiv \delta_3 + lfs_{t-1} - \beta_{34}lcp_{t-1} + \gamma_3 t$, and

$$\Delta lfd_t = a_2 + \Sigma b_{1k}\Delta lqp_{-k} + \Sigma b_{2k}\Delta lfd_{-k} + \Sigma b_{3k}\Delta lfs_{-k} + \Sigma b_{4k}\Delta lcp_{-k}$$
$$+ \alpha_{21}(CR1) + \alpha_{22}(CR2) + \alpha_{23}(CR3), \tag{7.9}$$

(Other equations with Δlfs or Δlcp as a dependent variable are omitted because they are not used in the following causality examination.) Here, *k* runs from 1 to an appropriate number which is dictated by the existence of causal relations and by the absence of serial correlation in Lagrange multiplier tests. The term *lfd* is one of the two financial development indicators, expressed in logarithms, and *lfs* is one of the two financial structure indicators, also in logarithms. Further, a_i, a_{ij}, b_{ij}, α_{ij}, β_{ij}, γ_i, and δ_i are all constants to be estimated. With Δ being a first difference, Δlfd, for example, stands for the growth rate of *fd*, where *fd* is the sum of the sizes of stock markets and banking markets (or money supply *m2*), each divided by (the GDP deflator × population); as was noted earlier, in the UK case, because of data availability, stocks traded are replaced by stocks issued in each year. A letter t in each cointegrated relation is a time trend (a number of years; we will return later on to the criteria for including t in cointegrated relations when we discuss causality patterns for each country and period).

Next, we examine if the meaningful cointegrations (with a negative and significant α_{ii}) have at least two unit root variables, so that error terms can be stationary. (This condition is a necessary, but not a sufficient, condition for the cointegrated relation to show some causality.) We conduct the unit root tests for all the variables . The tests use a method developed by Kwiatkowski et al. (1992; KPSS), as well as the augmented Dickey-Fuller tests (ADF). The relative powerfulness between the two tests is still in a controversial stage; see Carrion-i-Silvestre and Sanso (2006) and Muller (2005) for a useful comparison of various tests. The KPSS test statistics and the 1, 5, and 10% significance levels (under the null that the variable is stationary) are shown in Appendix B(1).[8] The test reveals that all the meaningful cointegrations showing causality, via cointegrated relations or loading factors, satisfy the above condition at the 1% or 5% significance level except for prewar UK, where the second and third cointegrations have two unit root variables at the 10% level; see Equation (7.4). In the ADF tests (with the null of there being a unit root), which are shown in Appendix B(2), for all the meaningful cointegrations, the existence of at least two unit-root variables cannot be rejected at the 10% level. Hence one can regard that the two tests yield quite consistent results.

One is now prepared to test, for each country and period, causalities between financial development and economic growth and also if changes in financial structure caused changes in the level or growth of per capita GDP.

7.3 Causality tests for individual countries and periods

We start with the prewar US period. As Table 7.1 exhibits, financial development measure *lsmp* yields one meaningful cointegrated relation, which implies that a general financial development causes an increase in per capita real output.[9] For another measure, *lscp*, one has two cointegrated relations, but only the first one is meaningful; this is also shown in Table 7.1. Appendix C, which is omitted from this chapter, provides the three criteria for whether the VAR system should contain a trend term (see note 4 in the previous chapter). The cointegrated relation for this period implies that *lscp* as well as financial structure measure *lsc* (a relative development of indirect compared with direct finance) raise per capita output. As Table 7.1 shows, the *p*-values in the Lagrange multiplier test (*plm*), with the null of no serial correlation, imply the residuals do not have serial correlations up to the specified lag number, so that they are (covariance) stationary. Also, *p*-values in the vector-error-correction White heteroskedasticity test (*pwh*), with the null of no heteroskedasticity, imply that the residuals are not heteroskedastic. Hence, the error terms are white noise, so that the estimated long-run relations can properly be called cointegrated relations. This statement can be applied to all the estimated long-run relations listed in Table 7.1.

Turning to the postwar US period, we have one cointegrated relation for financial development indicator *lsmp*, and another cointegration for indicator *lscp*. Only the former one cointegration case is shown in the table because this case presents wider causality patterns. As the table shows, the relationship implies that a general financial development as well as a relative development of direct compared with indirect finance, stimulate real-side development (*lqp*).[10]

Next, the prewar UK period has shown the causal patterns summarized in the table. Here, although the data on bank lendings are not available, bank deposit holdings generate three cointegrated relations, all of which are meaningful. From the first cointegration and loading factors, one can detect causalities from overall financial development to output as well as financial structure to output. Also, Equation (7.8) and the sign of α_{13} (<0) imply that the relative development of indirect compared to direct finance prompts growth of per capita income. Appendix D describes how the other three plausible cointegration matrices for $r = 3$ generate fewer causality patterns than when (7.4) is used.

In the postwar UK period, one meaningful cointegrated relation was found, as the table presents, which says that total financial development as well as indirect relative to direct financial development led to increases in per capita output. An alternative cointegrated relation with total financial development (*lsmp*) as a dependent variable does not have a meaningful (that is, a negative and significant) loading factor α_{11}.

Table 7.1 Cointegrations, loading factors, and causalities

Prewar USA ($obs = 53$, $co = 1$, $lag = 5$, $plm = 0.611$, $pwh=0.370$):

$lqp = 0.197lsmp - 0.021lsm + 0.171lcp + 0.009t$
        (2.737)   (0.565)   (4.758)   (4.247)
$\alpha_{11} = -1.440$ (4.392).

($obs = 53$, $co = 2$, $lag = 4$, $plm = 0.789$, $pwh = 0.813$):

$lqp = 0.353lscp - 0.129lsc + 0.008t$
        (16.894)   (15.749)   (17.279)
$\alpha_{11} = -0.715$, $\alpha_{21} = 0.174$, $\alpha_{12} = -0.104$, $\alpha_{22} = -0.043$.
        (2.016)   (0.177)   (1.629)   (0.244)

Postwar USA ($obs = 51$, $co = 1$, $lag = 5$, $plm = 0.286$, $pwh = 0.358$):

$lqp = 0.029lsmp + 0.026lsm + 0.287lcp + 0.011t$
        (2.240)   (2.480)   (12.545)   (9.330)
$\alpha_{11} = -0.616$ (4.545).

Prewar UK ($obs = 46$, $co = 3$, $lag = 5$, $plm = 0.062$, $pwh = 0.290$):

$lqp = 0.614lsdp - 0.010t$ (first cointegration)
        (18.712)   (3.199)
$\alpha_{11} = -1.098$ (2.319), $\alpha_{21} = 1.867$ (1.585), $\alpha_{31} = 3.173$ (0.667),
$\alpha_{12} = -0.104$ (1.833), $\alpha_{22} = -0.655$ (4.640), $\alpha_{32} = -1.506$ (2.640),
$\alpha_{13} = -0.358$ (2.521), $\alpha_{23} = -1.672$ (4.728), $\alpha_{33} = -3.868$ (2.707).

Postwar UK ($obs = 48$, $co = 1$, $lag = 5$, $lmp = 0.398$, $pwh = 0.351$):

$lqp = 0.052lsmp - 0.188lsm + 0.620lcp$ (no trend)
        (2.678)   (10.516)   (20.898)
$\alpha_{11} = -0.322$.

Prewar Japan ($obs = 42$, $co = 2$, $lag = 5$, $plm = 0.559$, $pwh = 0.186$):

$lqp = 4.429lscp - 3.559lsc - 0.277t$ (first cointegration)
        (141.865)   (4.056)   (14.361)
$\alpha_{11} = -0.434$ (1.985), $\alpha_{21} = -1.004$ (1.287),
$\alpha_{12} = -2.115$ (2.106), $\alpha_{22} = -4.450$ (1.242).

Postwar Japan ($obs = 46$, $co = 3$, $lag = 6$, $plm = 0.640$, $pwh = 0.266$):

$lqp = 0.882lsmp - 0.016t$ (first cointegration)
        (9.808)   (2.673)
$\alpha_{11} = -0.288$ (2.769), $\alpha_{21} = -0.883$ (1.746), $\alpha_{31} = 2.387$ (1.112),
$\alpha_{12} = 0.291$ (2.813), $\alpha_{22} = 0.378$ (0.749), $\alpha_{23} = 3.092$ (1.445),
$\alpha_{13} = -0.126$ (2.138), $\alpha_{23} = 0.041$ (0.144), $\alpha_{33} = 0.969$ (0.796).

Notes: (1) *obs*: the effective observation number. (2) *co*: the number of cointegration. (3) *lag*: lag order of the VAR system. (4) *plm*: the p-value in the LM test with the specified lag number, under the null that the VAR does not have residual autocorrelation. (5) *pwh*: the p-value in the vector-error-correction White heteroskedasticity test (without cross terms), with the null of no heteroskedasticity in error terms. (6) The estimated second cointegration when the cointegration number is two, and the last two cointegrations when the cointegration number is three, are not spelled out because they are not used in the inference of causality. (7) Values in parentheses after (or under) the estimated a_{ij}s are t-ratios in absolute value. (8) Italicized coefficients are the ones relevant to causality detection. (9) The critical value in the t-distribution for one-sided 5% significance level is 1.676 when the sample is 50.

Prewar Japan has two cointegrations, of which the first one is meaningful. The table presents the meaningful cointegrated relation and loading factors that are relevant to our causality inquiry. The cointegrated relation implies that overall financial development, as well as indirect relative to direct financial development, led Japan to higher per capita income. The only information that the loading factors provide is that the first, but not the second, cointegration is meaningful.

Examination on postwar Japan completes our causality detection. Here, for financial development measure *lsmp*, one has three cointegrated relations, but only the first one has a negative and significant loading factor. This is shown in the table. The first cointegration implies the causality running from overall financial development to per capita output. Also, the loading factor in this cointegrated relation α_{21} has the t-ratio $= 1.746$, whose p-value for one-sided test $= 0.044$ (the sign of α_{21} is not relevant so that a one-sided test can be applied here). Hence one concludes that causality runs also from per capita output to financial development (that is, the causality from the first to the second variable exists). The causal directions when the other three cointegration matrices are used are fewer than (and included in) those when the matrix (7.4) is used; see Appendix D for more discussion. In summary, one can conclude that there is a bidirectional causality between financial development and real-side development, but changes in financial structure had no influence on per capita output.

Table 7.2 sums up the causal directions and how changes in financial structure affected the growth rate of per capita GDP for each country

Table 7.2   A summary of causal directions

	output ↔ fin.level?	output ← fin.struc?	how it affects
Prewar USA:	← (c)	← (c)	indir.fin ↑→ output ↑
Postwar USA:	← (m)	← (m)	dir.fin ↑→ output ↑
Prewar UK:	← (d)	← (d)	indir.fin ↑→ output ↑
Postwar UK	← (m)	← (m)	indir.fin ↑→ output ↑
Prewar Jpn:	← (c)	← (c)	indir.fin ↑→ output ↑
Postwar Jpn:	↔ (m)	none	none

Notes: (1) Arrows represent causal directions. (2) '↔' shows bidirectional causality. (3) The letters after arrows represent the financial variables for which the causality appears, where *m*: *m2* money supply, *c*: bank credits, *d*: bank deposits. (4) The last column shows how the change in financial structure affects per capita output; for example, 'indir.fin ↑→ output ↑' means that an increase in indirect finance relative to direct finance expands per capita output.

and period. Also, Appendix E computes weak exogeneity tests and Toda and Phillips non-causality tests using χ^2 statistics, and endorses the results so far obtained.

7.4 Conclusions

This chapter has examined the causal patterns between per capita output and total financial development, and whether changes in financial structure caused (the speed of) output expansion.

A new methodological device of this chapter is that, applying the statistical analyses of Wickens (1996) and Pesaran and Shin (2002), we were able to set up economically plausible and statistically significant sets of (long-run) cointegrated relationships and to choose a set that exhibits the most comprehensive causality patterns, when the number of cointegrations is either two or three. This chapter also examined short-run causalities arising from significant loading factors α_{ij}, which are added to the causalities arising from cointegrated relations (see Equations (7.8) and (7.9)).

As for the causality between output and overall financial development, it was suggested that the Gurley-Shaw hypothesis was broadly relevant, where the hypothesis implies that causality runs from financial development to real-side development. An exception occurs in postwar Japan where causality is bidirectional.

Also, we derived the causal patterns from financial structure to per capita output in five cases. That is, except for postwar Japan, where we could not find the effect of change in financial structure on output, the other five periods saw that effect. While in postwar USA the change in financial structure toward proportionally larger direct finance (market finance) made for higher per capita output, in the other four cases, indirect finance (bank finance) was more favorable for larger per capita output.

Two interesting observations are that, in four cases out of the total six, overall financial development was a causal (not a resulting) factor for increases in per capita output, and that, also in four cases, relative expansion of indirect finance was more effective than direct finance in raising per capita output.

We presume that economic explanations for these results based on more or less detailed historical knowledge would be an interesting and complementary step to be taken. However, as these tasks belong to an independent research area, they have to be relegated to another occasion.

Finally, a word of caution is in order here. The present and previous chapter deal with causal directions between per capita output and financial development, with different choices of variables. In the USA and Japan, the directions are not very different for prewar and postwar periods in this and the previous chapter, but in the UK they are opposite between the two. This is presumably due to the fact that, notwithstanding the observations in note 2 of this chapter, for the UK, we had to use the values of new stock issues as a proxy for stock sales, because, for the UK, the latter series was not available, at least from the two volumes on British historical statistics, which are regarded as the best documents (see the Data Sources below). In any case, as for the output-financial development nexus, we will have to regard the results in Chapter 6 as more important, because the system in Chapter 6 includes two exogenous demand factors which might possibly affect output growth, and because the system in Chapter 7 involves a financial structure variable, which has less to do with the output-finance causality.

List of Symbols

lqp   per capita real GDP in logarithm.

m2   money supply consisting of cash in circulation, and demand and time deposits. The UK's *m3* is called here *m2*.

lscp   ln[(stock sales (issues) + bank credits)/(price level × size of population)], where bank credits are those to private sectors, and the numerator represents the total finance.

lsmp   above RHS (right-hand side) with bank credits replaced by *m2*.

lsdp   above RHS with *m2* replaced by bank deposits.

lsc   ln[stock sales (issues)/bank credits], where stock sales (issues) represent direct finance, and bank credits indirect finance.

lsm   above RHS with bank credits replaced by *m2*.

lsd   above RHS with *m2* replaced by bank deposits.

lcp   ln(real capital formation/size of population), where capital formation includes that of the government.

lfd   running symbol implying one of *lscp*, *lsmp*, or *lsdp*, where *fd* stands for 'financial development.'

lfs   running symbol implying one of *lsc*, *lsm*, or *lsd*, where *fs* stands for 'financial structure.'

α_{ij} coefficient in error-correction mechanisms.

β_{ij} coefficient in cointegrated relations.

plm   *p*-value in the Lagrange multiplier test under the null hypothesis that the VAR does not have serially autocorrelated residuals.

pwh   *p*-value in the vector-error-correction White heteroskedasticity test, with the null of no heteroskedasticity.

LL   log likelihood.

AIC   Akaike information criterion.

BIC   Schwartz Bayesian information criterion.

Data Sources

The USA

Gordon, R. J. (1986) *The American Business Cycle: Continuity and Change*. Chicago: University of Chicago Press.

International Monetary Fund (various years) *International Financial Statistics*. Washinton D.C.

US Department of Commerce (1975) *Historical Statistics of the United States*. Washington D.C.

The Bank of Japan (various years) *Comparative Economic and Financial Statistics*.

The UK

Capie, F. and Webber, A. (1985) *A Monetary History of the United Kingdom, 1870–1982*. London: Allen and Unwin.

International Monetary Fund (various years) *International Financial Statistics*. Washington D.C.

Mitchell, B. R. (1988) *British Historical Statistics*. Cambridge: Cambridge University Press.

The Bank of Japan (various years) *Comparative Economic and Financial Statistics*.

Japan

Fujino, S. (1994) *Money Supply in Japan*. Tokyo: Keiso-shobo.

Hitotsubashi University Institute of Economic Research. *Long-Term Economic Statistics*, 1: Ohkawa, K. et al. (1974) *National Income*; 4: Emi, K. (1971) *Capital Formation*. Tokyo: Toyokeizai-shinposha.

Management and Coordination Agency (1988) *The Historical Statistics of Japan*. Japan Statistical Association.

The Bank of Japan (various years) *Economic Statistics Annuals*.

Notes

1. For the UK only, the sales of stocks are represented by new stock issues in each year; see Data Sources. The UK's money supply aggregate $m3$ corresponds to $m2$ in the other two countries. Hence, the UK's $m3$ is called $m2$ in this chapter.

2. For each country, s is a flow variable but c is a stock variable. Here, however, we interpret c as a service flow provided by a stock c, representing a measure of indirect financial development, just as K in the production function represents a service flow delivered by a stock K. We adopt this interpretation for the following reason. Postwar US's average of stock sales/credits ratio sc (a third variable of our VAR system) is 0.564, but the US's ratio of stock market capitalization (a stock variable)/credits is 2.485 (for 1998, derived from *Comparative Economic and Financial Statistics*, 2000, Bank of Japan, pp. 43 and 50; the source is the same for the other two countries). The corresponding two figures for the UK are 0.053 and 1.404; and for Japan, 0.333 and 0.814. Comparing these figures, the first figures are much closer to the observations made by Mayer (1990) that in no countries do companies raise a substantial amount of finance from securities markets; and that banks are the dominant source of external finance in all countries (Mayer 1990, p. 313).

   For more discussion and similar measures of financial structure, see Beck et al. (2001, pp. 208–10) in the context of cross-country growth regression.

3. Demirguc-Kunt and Levine (2001) and Rajan and Zingale (1998) also focus on stock markets to represent direct finance (or market finance) in their cross-country growth regressions.

4. Individual values for two periods of the three countries shown in Appendices A through E are provided by EViews 6 (Quantitative Micro Software, Irvine CA, 2007). All the computations reported in this paper were done with this software package.

5. As the above derivation shows, however, one cannot restrict two cointegration vectors uniquely (that is, one can change the positions of zeros and βs without affecting the rank of cointegration matrix). But for prewar

USA and Japan, where the number of cointegrations is at most two, one can easily confirm that causality results due to other cointegration matrices $[1 - \beta_{12} - \beta_{13} \, 0; \, -\beta_{21} \, 1 \, 0 \, -\beta_{24}]$ or $[1 - \beta_{12} \, 0 \, -\beta_{14}; \, 0 \, 1 \, -\beta_{23} \, -\beta_{24}]$ (the first four elements of each matrix are the first row of a 2×3 matrix) are included in the results that arise from the matrix (7.1); that is, the results from matrix (7.1) are the most comprehensive.

Also, if the four variables are ordered as $[lqp \; lfs \; lfd \; lcp]$ but the cointegration matrix (7.1) remains the same, one might expect that the causality results are different, where *fs* is a running symbol meaning either *scp* or *smp*, and *fd* meaning either *sc* or *sm*. However, in all the cases where the cointegration number is at most two, the two matrices yield exactly the same causality patterns. The main reason for this is probably that the first cointegrated relation between output and the two financial measures is the same between the two cases. See the next section where individual periods will be examined.

6. It turns out, however, that whether $\beta_{23} = 0$ or $\beta_{23} \neq 0$ does not affect the causality patterns among the first three variables of the VAR system.

7. Appendix D explains that the three alternative, economically plausible, cointegration matrices for $r = 3$, $[1 - \beta_{12} \, 0 \, 0; \, 0 \, 1 - \beta_{23} \, 0; \, 0 \, -\beta_{32} \, 1 \, 0]$, and so on, where, for example, the first (second) four elements are the first (second) row vector of the 3×4 matrix, yields the same or fewer causality patterns. See the following individual examinations of prewar UK and postwar Japan.

8. Appendix B(1, 2) are not attached to this chapter but are available from the author upon request. See note 4 in Chapter 6.

9. It can be shown that a cointegrated relation where *lsmp* is a normalized (dependent) variable does not have a negative and significant α_{11}.

10. When the financial development is measured by *lscp*, one cointegration emerges, but for this case, only the financial structure measure *lsc* affects *lqp* positively, while the *t*-ratio for the coefficient of overall financial development is not significant ($t = 1.454$), so that the relationship is not shown in the text.

References

Barro, R.J. (1991) Economic growth in a cross-section of countries. *Quarterly Journal of Economics* 106, 407–43.

Beck, T., A. Demirguc-Kunt, R. Levine, and V. Maksimovic (2001) Financial structure and economic development: firm, industry, and country evidence. In: A. Demirguc-Kunt and R. Levine, eds. *Financial Structure and Economic Growth*, pp. 189–241. Cambridge, MA: MIT Press.

Beck, T. and R. Levine (2002) Industry growth and capital allocation: does having a market- or bank-based system matter? *Journal of Financial Economics* 64, 147–80.

Carrion-i-Silvestre, J.L. and A. Sanso (2006) A guide to the computation of stationarity tests. *Empirical Economics* 31, 433–48.

Cheung, Y.-W. and K.S. Lai (1993) Finite-sample sizes of Johansen's likelihood ratio tests for cointegration. *Oxford Bulletin of Economics and Statistics* 55, 313–28.

Demetriades, P.O. and K. Hussein (1996) Does financial development cause economic growth? Time series evidence from 16 countries. *Journal of Development Economics* 51, 387–411.

Demirguc-Kunt, A. and R. Levine, eds. (2001) *Financial Structure and Economic Growth: A Cross-Country Comparison of Banks, Markets, and Development.* Cambridge, MA: MIT Press.

Goldsmith, R.W. (1969) *Financial Structure and Development.* New Haven: Yale University Press,.

Gurley, J.G. and E.S. Shaw (1960) *Money in a Theory of Finance.* Washington, D.C.: The Brookings Institution.

Hall, S.G. and A. Milne (1994) The relevance of *P*-star analysis to UK monetary policy. *Economic Journal* 104, 597–604.

King, R.G. and R. Levine (1993a) Finance and growth: Schumpeter might be right. *Quarterly Journal of Economics* 108, 717–38.

—— (1993b) Finance, entrepreneurship, and growth. *Journal of Monetary Economics* 32, 513–42.

Kormendi, R.G. and P.G. Meguire (1985) Macroeconomic determinants of growth: cross-country evidence. *Journal of Monetary Economics* 16, 141–63.

Kwiatkowski, D., P.C.B. Phillips, P. Schmidt, and Y. Shin (1992) Testing the null hypothesis of stationarity against the alternative of a unit root. *Journal of Econometrics* 54, 159–78.

Levine, R. and S. Zervos (1998) Stock markets, banks, and economic growth. *American Economic Review* 88, 537–58.

Lucas, R.E., Jr. (1988) On the mechanics of economic development. *Journal of Monetary Economics* 22, 3–42.

Luintel, K.B. and M. Kahn (1999) A quantitative reassessment of the finance-growth nexus: evidence from a multivariate VAR. *Journal of Development Economics* 60, 381–405.

Luintel, K.B., M. Khan, P. Arestis, and K. Theodoridis (2008) Financial structure and economic growth. *Journal of Development Economics* 86, 181–200.

Lutkepohl, H. (2005) *New Introduction to Multiple Time Series Analysis.* New York: Springer.

MacKinnon, J.G., A.A. Haug, and L. Michelis (1999) Numerical distribution functions of likelihood ratio tests for cointegration. *Journal of Applied Econometrics* 14, 563–77.

Mayer, C. (1990) Financial systems, corporate finance, and economic development. In: R.G. Hubbard, ed. *Asymmetric Information, Corporate Finance, and Investment*, pp. 307–32. Chicago: University of Chicago Press.

McKinnon, R.I. (1973) *Money and Capital in Economic Development.* Washington D.C.: The Brookings Institution.

Muller, UK (2005) Size and power of tests of stationarity in highly autocorrelated time series. *Journal of Econometrics* 128, 195–213.

Neusser, K. and M. Kugler (1998) Manufacturing growth and financial development: evidence from OECD countries. *Review of Economics and Statistics* 80, 638–46.

Pesaran, M.H. and Y. Shin (2002) Long-run structural modeling. *Econometric Reviews* 21, 49–87.

Rajan, R.G. and L. Zingale (1998) Financial dependence and growth. *American Economic Review* 88, 559–86.

Robinson, J. (1952) *The Rate of Interest and Other Essays*. London: Macmillan.

Toda, H.Y. and P. Phillips (1993) Vector autoregression and causality. *Econometrica* 61, 1367–93.

Wickens, M.R. (1996) Interpreting cointegrating vectors and common stochastic trends. *Journal of Econometrics* 74, 255–71.

Appendix A. Cointegration Rank Tests: Trace Statistics and Their 5% Critical Values

		Trace statistics	5% critical values
Prewar US:	$r = 0^*$	171.654	63.876
	$r \leq 1^*$	44.032	42.915
	$r \leq 2$	16.831	25.872
Postwar US:	$r = 0^*$	148.060	63.876
	$r \leq 1$	36.877	42.915
Prewar UK:	$r = 0^*$	99.136	63.876
	$r \leq 1^*$	48.968	42.915
	$r \leq 2$	23.379	25.872
Postwar UK:	$r = 0^*$	82.584	63.876
	$r \leq 1$	39.038	42.915
Prewar Jpn:	$r = 0^*$	115.439	63.876
	$r \leq 1^*$	60.415	42.915
	$r \leq 2$	22.739	25.872
Postwar Jpn:	$r = 0^*$	94.003	63.876
	$r \leq 1^*$	51.286	42.915
	$r \leq 2^*$	26.105	25.872
	$r \leq 3$	10.090	12.518

H_0 (Null hypothesis): rank (the number of cointegration) = r.
*: Rejection of H_0 at the 5% level.

The following three Appendices are omitted (see note 4 in Chapter 6):

Appendix B(1). Kwiatkowski-Phillips-Schmidt-Shin (KPSS) Tests for Unit Roots

Appendix B(2). Augmented Dickey-Fuller Tests for Unit Roots

Appendix C. The Three Criteria for Setting Trends in Cointegrated Relations

Appendix D. Causalities from Other Restrictions When the Number of Cointegration is Three

Here, we show that for prewar UK where the cointegrated relations were indicated to be three, other restrictions on the cointegrations yield the same or fewer causality results compared to the results from the restrictions in the text. Recall that the four variables are ordered as $[lqp\ lfd\ lfs\ lcp]$, where lfd and lfs are the running symbols for logged overall financial development and logged financial structure, respectively.

When the restrictions are such that $[1 - \beta_{12}\ 0\ 0;\ 0\ 1\ 0\ -\beta_{24};\ 0\ -\beta_{32}\ 1\ 0]$, where, for example, the first four elements of this vector represent the first row of 3×4 (cointegration) matrix, the VAR yields, for financial variable $lsdp$, $lag = 5$, with a trend in each equation, two meaningful cointegrations. The first of them concerns the causality, which is shown below along with the loading factors:

$$lqp = 0.614 lsdp - 0.010t, \quad plm = 0.062,$$
$$\quad\ (6.692)\qquad\ \ (3.064)$$

$$\alpha_{11} = -1.098\ (2.319),\ \alpha_{21} = 1.867\ (1.585),\ \alpha_{31} = 3.173\ (0.667),$$

$$\alpha_{12} = -0.141\ (2.521),\ \alpha_{22} = -0.658\ (4.728),\ \alpha_{32} = -1.521\ (2.707),$$

$$\alpha_{13} = -0.094\ (3.439),\ \alpha_{23} = -0.006\ (0.088),\ \alpha_{33} = -0.038\ (0.139).$$

These relations show only the causality from overall financial development to output, which is included in the case examined in the text.

Also, it will be easy to confirm that each of the other plausible identification matrices, $[1\ 0\ 0\ -\beta_{14};\ -\beta_{21}\ 1\ 0\ 0;\ 0\ -\beta_{32}\ 0\ 1]$ and $[1\ 0\ 0\ -\beta_{14};\ 0\ 1\ 0\ -\beta_{24};\ 0\ -\beta_{32}\ 0\ 1]$, yield one-way causality running from financial development to output.

In postwar Japan, where the cointegration number was indicated to be three, other plausible restrictions on the cointegrations yield the same or fewer causality results compared to the results from the restrictions in the text.

When the restrictions are such that $[1 - \beta_{12}\ 0\ 0;\ 0\ 1\ 0\ -\beta_{24};\ 0\ -\beta_{32}\ 1\ 0]$, the VAR yields, for financial variable $lsmp$, $lag = 6$, with a trend in each equation, only one meaningful cointegration:

$$lqp = 0.882 lsmp - 0.016t\ \text{(first cointegration)}, \quad plm = 0.640,$$
$$\quad (9.808)\qquad\ \ (2.673),$$

$$\alpha_{11} = -0.288\ (2.769),\ \alpha_{21} = -0.883\ (1.746),\ \alpha_{31} = 2.387\ (1.112),$$

$$\alpha_{12} = 0.291\ (2.813),\ \alpha_{22} = 0.378\ (0.749),\ \alpha_{23} = 3.092\ (1.445),$$

$$\alpha_{13} = -0.126\ (2.138),\ \alpha_{23} = 0.041\ (0.144),\ \alpha_{33} = 0.969\ (0.796).$$

From the above, one obtains bidirectional causality between *lqp* and *lsmp*, exactly as in the text.

When the cointegration matrix takes the form: [1 0 0 $-\beta_{14}$; $-\beta_{21}$ 1 0 0; 0 $-\beta_{32}$ 1 0], all the three cointegrations are not meaningful, so that one cannot detect any causality. Finally, when the matrix has the form: [1 0 0 $-\beta_{14}$; 0 1 0 $-\beta_{24}$; 0 $-\beta_{32}$ 1 0], only the first cointegration is meaningful, but one finds only one causality running from total financial development to per capita output.

In this examination we did not use a matrix whose first row is [1 0 $-\beta_{13}$ 0], because this implies that, in the long-run, financial structure affects per capita output but total financial development does not, which seems to be a peculiar assumption.

Appendix E. Other Causality Tests Confirming the Results Shown in Table 7.1

	Weak exogeneity $\chi^2(1)$ test (the null:$\alpha_{12} = 0$ or $\alpha_{13} = 0$)		Toda-Phillips $\chi^2(2)$ test (the null:$\alpha_{12}\beta_{12} = 0$)	
	$\chi^2 = t^2$	signif. level	$\alpha_{12}\beta_{12}$	signif. level
Prewar USA				
first coint ($\alpha_{12} = 0$)	7.491	0.007	–	–
second coint ($\alpha_{13} = 0$)	285.6	0.000	27.52	0.000
Postwar USA				
($\alpha_{12} = 0$)	5.018	0.025	–	–
Prewar UK				
($\alpha_{12} = 0$)	5.378	0.018	34.299	0.000
($\alpha_{13} = 0$)	6.355	0.012	–	–
Postwar UK				
($\alpha_{12} = 0$)	7.172	0.008	–	–
($\alpha_{13} = 0$)	110.586	0.000	–	–
Prewar Japan				
($\alpha_{12} = 0$)	$201*10^2$	0.000	298.78	0.000
Postwar Japan				
($\alpha_{12} = 0$)	96.200	0.000	27.590	0.000

Notes: (1) When the number of cointegration is one, there is no loading factor α_{12} or α_{13}, so that one has no Toda-Phillips test. (2) Toda and Phillips (1993) use Wald tests for their non-causality tests; see their equations (17) and (20), but since Wald statistics and χ^2 have the same asymptotic property, and the latter is much easier to compute, we here employ $\chi^2(2)$ tests, where the degree of freedom is two, because they are the joint tests for loading factors and cointegration coefficients. (3) When the cointegration number is three, we set up the cointegration matrix whose $\beta_{13} = 0$ (see Equation (7.4) of this chapter), hence, in the above table there are many periods when α_{13} does not appear.

Part IV

8

Has Growth Been Led by Investment or Exports? Prewar and Postwar US, UK, and Japan

8.1 Introduction

The recent few decades have seen a large volume of literature on econometric, cross-country quests for factors promoting growth rates of (per capita) real GDP in a wide range of countries. Though the empirical research is still going on, it is now widely recognized that domestic investment and the volume of exports (both usually in per capita terms or as proportions of GDP) are the two most robust factors promoting economic growth, irrespective of other variables included as possible candidates for growth promoters (Levine and Renelt (1992); some may assert that it will be a sum of exports and imports that affects a country's economic growth; however, we are concerned with two major *demand* factors that may affect economic growth; and also Levin and Renelt (1992, p. 953) note in their cross-country regressions that the results are essentially the same whether one uses exports, imports, or their sum, as the trade volume). Hence it will be a natural development in research agenda to inquire which of the two growth promoters, exports or investment, was more forceful in raising the growth rate in historical contexts of various countries.

As for the positive role of more active exports in economic growth, there has been a large number of empirical work, an earlier one being Feder (1982), and more recent examples being Devereux (1997), Krueger (1997), and Riviera-Batiz and Romer (1991). The growth-enhancing role of investment, on the other hand, has been recognized in the long history of growth economics. Here we only mention DeLong and Summers (1991) and Barro and Sala-i-Martin (2004).

Another group of work deals with causality directions between exports and economic growth and between investment and growth for some developing and developed countries; see, for example, Armad and Harnhirun (1995), Madsen (2002), and Yamada (1998), which use mostly time series analysis. The above time series analyses yield diverse results on causal directions and on the chief growth promoters, depending on the methods they employ and on the countries and periods they chose, although those attempts seem to be heading in a desirable direction.

In this chapter we develop some examination of causality that might run from exporting activity to growth as well as from domestic investment to growth, using vector-autoregression analyses with cointegration and error-correction systems. The method, developed by, for example, Hall and Milne (1994), Luintel and Kahn (1999), and Arestis, Demetriades and Luintel (2001), is of a relatively recent vintage and can estimate the coefficients in cointegrations and those in error-corrections simultaneously and therefore efficiently. Hence it may be called a generalized (or an extended) Granger causality test.

A new methodological device of this chapter is that it applies the formal results proposed by Wickens (1996), and Pesaran and Shin (2002), which maintain that for the identification of cointegration coefficients, one needs *a priori* economic theory-based restrictions on the coefficients. As is now well recognized, the (vector of) cointegration coefficients cannot be determined uniquely through statistical methods only, hence we examine a few (exhaustive) economically plausible sets of coefficients, and then adopt a representative set that generates the most comprehensive causality results.

The main focus of this chapter is on which of the above two variables, exports or investment, was more responsible for the growth of per capita GDP. The subjects of the analysis are prewar and postwar USA, UK, and Japan. We chose these three countries because of easier access to historical data with relatively fine quality. The four variables to be included in the VAR systems are: per capita real GDP, per capita real exports, per capita real capital formation, and one of two variables measuring financial development. The last variable was included as an extra demand factor to exports and investment that is likely to enhance the growth capacity of the countries; in other words, we control for financial and monetary environments that may influence economic growth.

The chapter is organized as follows. Section 2 describes the VAR framework that is used in the causality tests. Section 3 develops the tests for the two periods, prewar and postwar, of the three countries, and compares the causal directions for those countries and periods. Section 4 summarizes and concludes the chapter.

8.2 The method: the VAR systems and causality

The VAR system we use for detecting causality directions involves four variables: per capita real GDP (*lqp*; the first *l* indicates that the variable is expressed in the natural logarithm, which also applies to other variables starting with *l*), per capita real exports (*lep*), per capita real capital formation (*lcp*) where capital formation includes that of the government, and an index of financial development (*lfd*), where *lfd* is either per capita real *m2* (*lmp*) or the *m2*/(base money) ratio (*ln*, that is, the logged *m2*-money multiplier).[1] We define *m2* as the broader money supply including time deposits. The main symbols appearing in this chapter are gathered and described in the List of Symbols section following the text. The section on data sources appears next.

The period divisions for the US/UK and Japan are different because of the times when the two groups started 'Modern Economic Growth' (Kuznets (1971)) and because of events which are exogenous to the interests of this chapter such as the World War and Great Depressions. The US and UK's former (prewar) period is thus set from 1874 through 1926, while Japan's counterpart is set from 1888 through 1940. The latter (postwar) period runs from 1953 through 2005 uniformly for the three countries.

In order for the four variables to form a cointegrating vector, there needs to be at least two variables with a unit root in the group of [*lqp lep lcp lmp* (or *ln*)]. The unit root tests are conducted here using the KPSS test (Kwiatkowski et al. (1992)) which has the null of stationarity of the variable in question, and which is known to be more powerful in rejecting the unit-root hypothesis than other conventional tests. The test results are exhibited in Appendix A, which is omitted from the chapter but is available from the author (see note 4, Chapter 4 for more information).

The maximum number of cointegrating vectors is decided in view of trace statistics, which are more robust than the method using the maximum eigenvalues (Cheung and Lai (1993)). There will be cases with one, two, or three cointegrating vectors according to the criterion. Appendix B, which is omitted here, exhibits the trace statistics and appropriate cointegration numbers at the 5 or 1% significance level for all cases.[2]

According to Pesaran and Shin (2002, pp. 54–6), in order that the cointegrated relations can be identified, r^2 restrictions on the coefficients must be distributed across the r different cointegrated vectors such that there are r different restrictions (including a normalizing restriction) for each of the r cointegrating vectors, where r is the number of cointegrated relations (= the rank of the cointegrated matrix). In the two cointegration case with four variables, each relation (row) normally has $n - r = 2$ unrestricted variables, where n (= 4) is the number of variables, and r^2 (= 4) is the total number of restrictions to be distributed evenly to the two cointegrated relations (see Wickens (1996), and Pesaran and Shin (2002, pp. 54–6 and 77)). These authors maintain that cointegrated relations cannot be identified by statistical arguments only, so that the identification requires *a priori* economic theory-based restrictions on the cointegrated relations.

When the trace statistics indicate that the number of cointegrated relations is at most two, we express the error-correction mechanisms for financial variable, for example, *lmp*, as

$$\begin{bmatrix} \Delta lqp_t \\ \Delta lep_t \\ \Delta lcp_t \\ \Delta lmp_t \end{bmatrix} = \begin{bmatrix} \alpha_{11} & \alpha_{12} \\ \alpha_{21} & \alpha_{22} \\ \alpha_{31} & \alpha_{32} \\ \alpha_{41} & \alpha_{42} \end{bmatrix} \begin{bmatrix} 1 & -\beta_{12} & -\beta_{13} & 0 \\ 0 & 1 & -\beta_{23} & -\beta_{24} \end{bmatrix} \begin{bmatrix} lqp_{t-1} \\ lep_{t-1} \\ lcp_{t-1} \\ lmp_{t-1} \end{bmatrix} . \quad (8.1)$$

Here the first cointegrated relationship is, from the last two matrices,

$$lqp_{t-1} = \beta_{12} lep_{t-1} + \beta_{13} lcp_{t-1},$$

and the second relation is

$$lep_{t-1} = \beta_{23} lcp_{t-1} + \beta_{24} lmp_{t-1}.$$

The two relations have economically well-defined meanings: The first relation implies that, in the long-run, changes in per capita output are caused by changes in per capita exports and per capita investment. Also, the second relation means that per capita exports depend on per capita capital formation as well as financial development.

In the above two relations, a constant and a time trend are omitted, but in all the cases a constant is present, and in most cases a trend is present (see next section). Appendix C, omitted from this chapter, provides the criteria for including a time trend.

We can show that the above two cointegration vectors can be derived from the two general forms of cointegrated relations, using some manipulations that keep the linear independence of the two relations. See Chapter 6, Section 2 for this demonstration.

Hence, in Appendix D, we examine the causality results when the matrix takes other forms, where it is shown that the matrix used in the text yields the broadest causality patterns. This is arguably so because in the first cointegration vector, the matrix in the text treats exports and investment on an 'equal footing' (that is, generally, $\beta_{12} \neq 0$ and $\beta_{13} \neq 0$) (see Appendix D).

In Equation (8.1), the α_{ij}s are the speeds of adjustment (loading factors) in the error-correction mechanisms. For the first cointegrated relation to be meaningful, one needs to have a negative and significant α_{11}; see Luintel and Kahn (1999, pp. 388–9). In what follows, 'a meaningful cointegration' is equivalent to 'a cointegration having a negative and significant α_{ii}'. In addition, if α_{21} is significant (irrespective of its sign), one can judge that the causality runs from per capita GDP (the 1st variable) to per capita exports (the 2nd variable); the same is true for α_{22} and α_{12}; see the above paper and Lutkepohl (2005, chs 6 and 7) for more discussion.

When the trace statistics indicate that there are at most three cointegrated relations, the right-hand matrix of error-correction coefficients $[\alpha_{ij}]$ is 3×4, and each row (each cointegration) should have $r = 3$ restrictions, one of which should be a normalization restriction. If some row has fewer restrictions (more βs), one cannot identify all the αs and βs. Also, the restrictions must be such that the cointegration matrix has rank $r = 3$. Then, the cointegrating vectors are, for example,

$$\begin{bmatrix} 1 & 0 & -\beta_{13} & 0 \\ 0 & 1 & -\beta_{23} & 0 \\ 0 & 0 & 1 & -\beta_{34} \end{bmatrix}. \tag{8.2}$$

In (8.2), similarly to the two cointegration case, $r^2 = 9$ restrictions are distributed evenly across the three cointegrated vectors, with three restrictions (one being a normalizing restriction) on each cointegrated vector.

The first row of (8.2) implies that larger per capita output is caused by larger per capita investment (the investment-multiplier relation). The second row implies that increases in investment prompt exports. The third row means that financial development increases investment activity. We will also use an alternative cointegration matrix for $r=3$ whose first row is replaced by $[1 \ -\beta_{12} \ 0 \ 0]$ ('output is raised more by exports than by investment') and compare the results. The second cointegration for $r=3$ is therefore,

$$\begin{bmatrix} 1 & -\beta_{12} & 0 & 0 \\ 0 & 1 & -\beta_{23} & 0 \\ 0 & 0 & 1 & -\beta_{34} \end{bmatrix}. \tag{8.3}$$

The above two alternative matrices for $r = 3$ are all that one should consider, because, given the first two rows (the first row has two variants), the (3, 4) element of the matrix should be non-zero (that is, $-\beta_{34} \neq 0$) which determines the form of the third row. The case that involves three cointegrated relations occurs only in prewar USA, where it is shown that only the matrix (8.3), not (8.2), generates a causality pattern (see below).

In the case of three cointegrated relations, the first two VAR equations with error-correction terms $(\alpha_{ij}[CRj], \ j = 1,2,3)$ can be written as

$$\Delta lqp_t = a_1 + \Sigma a_{1k}\Delta lqp_{-k} + \Sigma a_{2k}\Delta lep_{-k} + \Sigma a_{3k}\Delta lcp_{-k} + \Sigma a_{4k}\Delta lfd_{-k}$$
$$+ \alpha_{11}(CR1) + \alpha_{12}(CR2) + \alpha_{13}(CR3),$$

where $CR1 \equiv \delta_1 + lqp_{t-1} - \beta_{13}lcp_{t-1} + \gamma_1 t$, $CR2 \equiv \delta_2 + lep_{t-1} - \beta_{23}lcp_{t-1} + \gamma_2 t$, $CR3 \equiv \delta_3 + lcp_{t-1} - \beta_{34}lfd_{t-1} + \gamma_3 t$, and

$$\Delta lep_t = a_2 + \Sigma b_{1k}\Delta lqp_{-k} + \Sigma b_{2k}\Delta lep_{-k} + \Sigma b_{3k}\Delta lcp_{-k} + \Sigma b_{4k}\Delta lfd_{-k}$$
$$+ \alpha_{21}(CR1) + \alpha_{22}(CR2) + \alpha_{23}(CR3).$$

(Other equations with Δlcp and Δlfd as a dependent variable are omitted because they are not relevant to our current interest.) Here, k runs from 1 to an appropriate number which is decided by the requirements that the VAR system shows some causality from exports to output or from investment to output as well as that the VAR system passes the Lagrange multiplier tests (LM tests) for serial correlation. In the above system, a_i, a_{ij}, b_{ij}, α_{ij}, β_{ij}, γ_i, and δ_i are all constants to be estimated, and t is a trend term (a number of years). For the two cointegration case, one has $\alpha_{13} \equiv \alpha_{23} \equiv 0$ with $CR1$ and $CR2$ appropriately modified (see the two relations following Equation (8.1)).

We have now seen all the instruments to examine whether the causality was extended from exports to output and/or from investment to output, for two periods of the three countries.

8.3 Causalities from exports to output or investment to output in the three countries

We start with the prewar US period. The trace statistics indicate that cointegration number 3 is significant at least at the 5% level. The first cointegration matrix (8.2) does not yield any causality. The matrix (8.3), on the other hand, generates a causality from exports to output, as the loading factors for this case exhibit. Only the second cointegrated relation is meaningful (α_{22} is negative and significant). Also α_{12} is significant, implying that a causality runs from exports (the second variable) to output (the first variable). Since the second cointegrated relation (a relation between investment and exports) does not involve any causality we are interested in, it is not shown in Table 8.1.

As the p-value of the Lagrange multiplier test for serial correlation (plm) implies, the residuals have no autocorrelations for lag 6; and also, as the p-value of the vector-error-correction White heteroskedasticity test (pwh) shows, error terms are not heteroskedastic, so that they are white noise and hence (covariance) stationary. This means that the second relation can be properly called a cointegrated relation; this is true for all the cointegrated relations shown in Table 8.1 as the above two p-values listed in the table indicate.

Turning to the postwar US period, we found one cointegration, which has a *negative* and significant coefficient on per capita exports.[3] But deleting *lep* to estimate a three-variable VAR yields one cointegration, which is shown in the table. The cointegration implies that there is a causality running from investment to output. Deleting *lcp* and restoring *lep*, however, did not generate any cointegration.

For the prewar UK period, we found two cointegrated relations, the first of which is meaningful as the loading factors α_{11} and α_{22} indicate. The first cointegration then says both exports and investment caused

Table 8.1　Cointegration, loading factors, and causality

Prewar USA (1874–1926; $ln,$[1] $l = 6,$[2]$co = 3,$[3]$plm = 0.094,$[4] $pwh = 0.467,$[5] a trend in each equation)

$\alpha_{11} = -0.396,\ \alpha_{21} = 1.343,\ \alpha_{31} = 0.945,$
$\quad (1.275)$[6]$\qquad (1.361) \qquad\quad (1.880)$

$\alpha_{12} = -0.259$[7]$,\ \alpha_{22} = -0.755,\ \alpha_{32} = 0.945,$
$\quad (2.117) \qquad\quad (1.943) \qquad\quad (2.446)$

$\alpha_{13} = 0.090,\ \alpha_{23} = 1.158,\ \alpha_{33} = 1.454.$
$\quad (0.225) \qquad\ (0.906) \qquad\quad (2.235)$

(Only the second cointegration is meaningful, that is, α_{22} is negative and significant.)

Postwar USA (1953–2005; $ln,$ $l = 5,$ $co = 1,$ $plm = 0.780,$ $pwh = 0.884,$ a trend in the equation)

$lqp = 0.174lcp + 0.059ln + 0.016t,$
$\quad\ \ (4.021) \qquad (2.742) \qquad (17.138)$
$\alpha_{11} = -0.87\ (3.526).$

Prewar UK (1874–1926; $lmp,$ $co = 2,$ $plm = 0.128,$ $pwh = 0.420,$ a trend in each equation)

$lqp = 0.719lep + 0.710lcp - 0.0002t,$
$\quad\ \ (3.347) \qquad (6.401) \qquad\ (0.098)$

$\alpha_{11} = -0.167,\ \alpha_{21} = -0.436,\ \alpha_{12} = -0.017,\ \alpha_{22} = 0.131.$
$\quad (1.942) \qquad\quad (2.745) \qquad\quad (0.190) \qquad\quad (0.813)$

(Only the first cointegration is meaningful.)

Postwar UK (1953–2005; $lmp,$ $co = 2,$ $plm = 0.461,$ $pwh = 0.293,$ no trend in any equation)

$\alpha_{11} = 0.059,\ \alpha_{21} = -0.401, \alpha_{31} = 0.012, \alpha_{12} = -0.044,\ \alpha_{22} = -0.173,\ \alpha_{32} = -0.067.$
$\quad (1.473) \qquad\quad (3.263) \qquad\ (0.110) \qquad\quad (2.349) \qquad\qquad (3.128) \qquad\qquad (1.302)$

Prewar Japan (1888–1940; $ln,$ $l = 6,$ $co = 2,$ $plm = 0.693,$ $pwh = 0.342,$ a trend in each equation)

$lqp = -0.229lep + 0.217lcp + 0.019t,$
$\qquad\ \ (4.867) \qquad\ (4.553) \qquad (7.999)$

$\alpha_{11} = -0.903,\ \alpha_{21} = -0.915,\ \alpha_{12} = 3.314,\ \alpha_{22} = 0.888.$
$\quad (3.314) \qquad\quad (0.723) \qquad\quad (1.908) \qquad\quad (0.741)$
(Only the first cointegration is meaningful.)

Postwar Japan (1953–2005; $ln,$ $l = 3,$ $co = 2,$ $plm = 0.172,$ $pwh = 0.201,$ a trend in each equation)

$\alpha_{11} = 0.014,\ \alpha_{21} = 1.091,\ \alpha_{12} = 0.051,\ \alpha_{22} = -0.123.$
$\quad (0.097) \qquad\ (2.128) \qquad\ (2.765) \qquad\quad (1.908)$

(Only the second cointegration is meaningful.)

Notes: (1) The financial variables after time periods are those for which causalities were detected. (2) l: a lag number. (3) co: a number of cointegration. (4) plm: a p-value in the LM test with the null hypothesis that there is no serial correlation in residuals. (5) pwh: a p-value in the White heteroskedasticity test for error terms, with the null of no heteroskedasticity. (6) Figures in parentheses are t-ratios in absolute value. (7) Italicized coefficients are those relevant to causality judgement.

Table 8.2   Causal directions: a recapitulation

Prewar USA: $lqp \leftarrow lep$ (ln).[1]	**Postwar USA:** $lqp \leftarrow lcp$ (lmp).
Prewar UK: $lqp \leftarrow lep$, lcp (both lmp).[2]	**Postwar UK :** $lqp \leftarrow lep$ (lmp).
Prewar Japan: $lqp \leftarrow lcp$ (ln).	**Postwar Japan:** $lqp \leftarrow lep$ (ln).

Notes: (1) '$lqp \leftarrow lep$' implies that causality exists from exports to per capita income. The variable in parentheses is a financial variable for which the causality was found. (2) The effects of lep and lcp on lqp are of the similar size because their coefficients are nearly equal in the cointegrated relation.

output. Also the coefficients show that the force of causation is about the same in each.

The postwar UK period provides two cointegrations, but only the second one is meaningful. This is not shown in the table, however, because it does not indicate any causal pattern we are interested in. Since α_{22} is negative and significant and α_{12} is significant, a causality from exports to output can be detected.

In the prewar Japanese period, we found only one meaningful cointegration, which, however, indicates that exports and domestic investment are substitute demand factors, with the negative coefficient on exports. Although this suggests that investment is a causal factor for output, we conducted traditional Granger causality (F) tests, which clearly support the above conjecture.[4]

It now remains to examine the postwar Japanese period. Two cointegrated relations were found, but only the second one is meaningful as the loading factors show. Also, a significant α_{12} implies that there is a causality running from exports to output. Table 8.2 summarizes the causality results which have been detected so far in this section.

8.4   Conclusions

Using four-variable VAR systems with cointegration and error-correction mechanisms, which consist of output, exports, domestic investment, and one of the two financial development indicators, this chapter tried to extract the causality patterns from exports to output or from investment to output. The exercises were conducted for prewar and postwar USA, UK, and Japan. In other words, the chapter's primary intention was to see whether the growth processes of these countries and periods were export-led or investment-led.

A new methodological device of this chapter is that, applying the formal analyses of Wickens (1996), and Pesaran and Shin (2002), it was possible to formulate economically plausible and statistically significant sets of cointegrated (long-run equilibrium) relationships and to choose a set that exhibits the most comprehensive causality results, when the cointegration number is either two or three.

The conclusions obtained are that in the USA, growth has changed from export-led to investment-led. Prewar UK's growth was promoted by both exports and investment, while its postwar counterpart was led mainly by exports. Finally, prewar Japan's growth was characterized as investment-led, but in the postwar period, it has changed to be export-led one.

We would add finally that the above conclusions may be better reinforced by arguments using some knowledge of relevant economic history. However, in view of a considerably common understanding that time series analysis is better suited for extracting plausible causal relationships than regression analysis, this chapter might be able to claim some contribution in empirical causality analyses.

List of Symbols

lqp per capita real output expressed in logarithm (the variables beginning with l are all expressed in the same manner).

lep per capita real exports.

lcp per capita real capital formation, which includes that of the government.

lmp per capita real amount of $m2$, where $m2$ consists of cash currency as well as demand and time deposits.

ln $m2$-money multiplier, that is, $m2$/base money.

lfd either lmp or ln.

t time trend (1874 = 1).

α_{ij} loading facor (speed of adjustment) in the error-correction mechanism.

β_{ij} coefficient in a cointegrated relation.

n number of variables in the VAR system.

r number of cointegrated relations (that is, the rank of the cointegration matrix).

Data Sources

The USA

Gordon, R.J. (1986) *The American Business Cycle: Continuity and Change.* Chicago: University of Chicago Press.

International Monetary Fund (various years) *International Financial Statistics.* Washington DC.

US Department of Commerce (1975) *Historical Statistics of the United States.* Washington DC.

The UK

Capie, F. and Webber A. (1985) *A Monetary History of the United Kingdom, 1870–1982.* London: Allen and Unwin.

International Monetary Fund (various years) *International Financial Statistics.* Washinton DC.

Mitchell, B.R. (1988) *British Historical Statistics.* Cambridge: Cambridge University Press.

Japan

Fujino, S. (1994) *Money Supply in Japan.* Tokyo:Keiso-shobo.

Hitotsubashi University Institute of Economic Research. *Long-Term Economic Statistics,* 1: Ohkawa, K. et al. (1974) *National income;* 4: Emi, K. (1971) *Capital Formation.* Tokyo: Toyokeizai.

Management and Coordination Agency (1988) *The Historical Statistics of Japan.* Japan Statistical Association.

The Bank of Japan (various years) *Economic Statistics Annuals.*

Notes

1. In some papers dealing with causality between growth of per capita income and financial development, the latter is also represented by commercial bank loans or banks' liabilities; see, for example, Arestis and Demetriades (1997), Luintel and Kahn (1999), and Arestis, Demetriades, and Luintel (2001). But the major proportions of the asset side and liability side of banks' balance sheets are composed of loans and deposits, respectively, which are nearly of equal size, and money is defined by the sum of deposits and proportionally much smaller cash in circulation. Hence money supply may be seen as at least as good and comprehensive a measure of financial development as loans or deposits.

   The output expanding effect of inside money (and, therefore, money multipliers) is noted by King and Plosser (1984).

2. The statistics which are exhibited in Appendices A, B, and C are computed by the econometric software package EViews, Ver. 6 (2007). All the computations in preparing this chapter were done with this package.

3. $lqp = -0.085lep + 0.232lcp + 0.131ln + 0.018t,$
     $\quad\quad (2.570) \quad\quad (4.796) \quad\quad (4.499) \quad (18.522)$

   where the figures below the coefficients are t-ratios in absolute value.

4. Let the sum of squared residual (SSR) from the OLS of lqp on $lep(-1)$, $lcp(-1)$, $ln(-1)$, a constant, and a trend be S_2. Let the SSR from the OLS when $lep(-1)$ ($lcp(-1)$) is dropped from the explanatory variables be S_0 (S_1). Then $S_2 = 0.023$, $S_0 = 0.026$, $S_1 = 0.036$.

   The F statistic regarding lep is $[(26-23)/1]/[23/(53-5)] = 6.26$, where the effective observation number is 53, explanatory variables (including a constant and a trend) are five, and the constraint number is one. Similary, for lcp, $F = [(36-23)/1]/[23/(53-5)] = 27.13$.

   The upper 1% (5%) point of the F-distribution with $df(1,48)$ is 7.21 (5.37), hence the null that lep's (lcp's) coefficient is zero cannot be (can be) rejected at the 1% level. One cannot, however, get the same conclusion when the right-hand side consists of the sum of lagged variables from two to five orders.

References

Arestis, P. and P.O. Demetriades (1997) Financial development and economic growth: assessing the evidence. *Economic Journal* 107, 783–99.

Arestis, P., P.O. Demetriades, and K.B. Luintel (2001). Financial development and economic growth: the role of stock markets. *Journal of Money, Credit, and Banking* 33, 16-41.

Armad, J. and S. Harnhirun (1995) Unit roots and cointegration in estimating causality between exports and economic growth: empirical evidence from the ASEAN countries. *Economics Letters* 49, 329–34.

Barro, R.J. and X. Sala-i-Martin (2004) *Economic Growth*, 2nd ed. Cambridge, Mass.: MIT Press.

Cheung, Y.-W. and K.S. Lai (1993) Finite-sample sizes of Johansen's likelihood ratio tests for cointegration. *Oxford Bulletin of Economics and Statistics* 55, 313–28.

DeLong, J.B. and L.H. Summers (1991) Equipment investment and economic growth. *Quarterly Journal of Economics* 106, 445–502.

Demirguc-Kunt, A. and R. Levine (2001) *Financial Structure and Economic Growth: A Cross-Country Comparison of Banks, Markets, and Development.* Cambridge, Mass.: MIT Press.

Devereux, M.B. (1997) Growth, specialization, and trade liberalization. *International Economic Review* 38, 565–85.

EViews, Ver. 6 (2007) Irvine, CA: Quantitative Micro Software.

Feder, G. (1982) On exports and economic growth. *Journal of Development Economics* 12, 59–72.

Hall, S.G. and A. Milne (1994) The relevance of *P-star* analysis to UK monetary policy. *Economic Journal* 104, 597–604.

King, R.G. and C.I. Plosser (1984) Money, credit, and prices in a real business cycle. *American Economic Review* 74, 363–80.

Krueger, A.O. (1997) Trade policy and economic development: how we learn. *American Economic Review* 87, 1–22.

Kuznets, S. (1971) *Economic Growth of Nations: Total Output and Production Structure.* Cambridge, Mass.: Harvard University Press.

Kwiatkowski, D., P. Phillips, P. Schmidt, and Y. Shin (1992) Testing the null hypothesis of stationarity against the alternative of a unit root. *Journal of Econometrics* 54, 159–78.

Levine, R. and D. Renelt (1992) A sensitivity analysis of cross-country growth regressions. *American Economic Review* 82, 942–63.

Luintel, K.B. and M. Khan (1999) A quantitative reassessment of finance-growth nexus: evidence from a multivariate VAR. *Journal of Development Economics* 60, 381–405.

Lutkepohl, H. (2005) *New Introduction to Multiple Time Series Analysis.* New York: Springer.

Madsen, J.B. (2002) The causality between investment and economic growth. *Economics Letters* 74, 157–63.

Pesaran, M.H. and Y. Shin (2002) Long-run structural modeling. *Econometric Reviews* 21, 49–87.

Riviera-Batiz, L. and P.M. Romer (1991) Economic integration and endogenous Growth. *Quarterly Journal of Economics* 56, 531–55.

Wickens, M.R. (1996) Interpreting cointegrating vectors and common stochastic trends. *Journal of Econometrics* 74, 255–71.

Yamada, H. (1998) A note on the causality between export and productivity: an empirical re-examination. *Economics Letters* 61, 111–14.

Appendices

The following Appendices A, B, and C are omitted from this chapter, but are available from the author (see note 4, ch 6).

Appendix A. Kwiatkowski-Phillips-Schmit-Shin (KPSS) Unit Root Tests

Appendix B. Cointegration Rank tests: Trace Statistics and Their 5% Critical Values

Appendix C. Three Tests for the Existence of a Trend in Cointegrated Relations

Appendix D. A Summary of Causalities That Arise from Other Cointegrated Relations When the Number of Relations is Two

Define the following cointegrated matrices which are alternative to the one used in the text (the middle matrix on the right-hand side of Equation (8.1), which is called P).

$A \equiv [1 \ 0 \ -\beta_{13} - \beta_{14}; \ 0 \ 1 \ -\beta_{23} - \beta_{24}]$, where the first (second) four elements are the first (second) row of a 2×4 matrix. Also,

$B \equiv [1 \ -\beta_{12} - \beta_{13} \ 0; \ -\beta_{21} \ 1 \ 0 \ -\beta_{24}]$.

The matrix $B' \equiv [1 \ -\beta_{12} \ 0 \ -\beta_{14}; \ 0 \ 1 \ -\beta_{23} - \beta_{24}]$ is omitted because this excludes , to start with, the causality running from investment to output.

Changing now the order of the four variables to [*lqp lcp lep lfd*], define

$C \equiv [1 \ -\beta_{12} \ -\beta_{13} \ 0; 0 \ 1 \ -\beta_{23} \ -\beta_{24}]$.

The upshot of the following arguments is that matrix P used in the text yields more comprehensive causality results than other matrices, A, B, and C.

Prewar UK

If matrix A is used (instead of P), a causality from investment to output appears, but the reverse causality does not.

If matrices B and C are used, causalities appear which are exactly the same as with P.

Postwar UK

For matrices A and B, one obtains the same causal pattern as with P. For matrix C, the two cointegrated relations are not meaningful, that is, either non-negative or insignificant, so that one cannot find any causality.

Prewar Japan

Matrix A yields the same causality pattern as with P, while matrix B yields no meaningful cointegrated relation. For matrix C one finds the same causal direction as with P.

Postwar Japan

Matrices A and B yield the same causality as for P, but for matrix C, the two cointegrations are not meaningful so that no causality is found.

9
Was It Investment or Exports That Led Economic Growth? 13 Developing Country Experiences

9.1 Introduction

In a paper which dealt with the strength of various explanatory variables in cross-country growth regressions, Levine and Renelt (1992) found the (domestic) investment-output ratio to be the most robust variable in explaining the growth of per capita output of developed and developing countries. They also found that the trade amount-output ratio usually affects the investment-output ratio. Combining these two points suggests that the next task would be to examine which is the more forceful variable, between investment and exports, in causing country's per capita output to grow faster.

Regarding the positive role of exports in promoting economic growth, there has been a large number of empirical works supporting this role, such as Feder (1982), Devereux (1997), Krueger (1997), and Riviera-Batiz and Romer (1991). The growth-enhancing role of investment has also been recognized in the long history of growth economics, with examples being DeLong and Summers (1991) and Barro and Sala-i-Martin (2004). Another group of work deals with causality directions between exports and economic growth, and between investment and growth for some developing and developed countries; see, for example, Armad and Harnhirun (1995), Madsen (2002), and Yamada (1998), which use mostly time series analysis. The results of the above time series analyses do not seem to have converged yet on the causal directions and on the chief growth promoters, depending on the methods they employ and on the countries and periods they chose. However, these attempts should be regarded as dealing with important and useful topics in empirical dynamics and policy analyses.

In this chapter we develop some examination of causality that might run from exporting activity to growth or from domestic investment to growth, using vector-autoregression analyses with cointegration and error-correction mechanisms. These methods belong to relatively new analytical tools, developed by Hall and Milne (1994), Arestis and Demetriades (1997), Luintel and Kahn (1999), and Arestis, Demetriades and Luintel (2001), which can estimate the coefficients in cointegrations and those in error-corrections simultaneously and, therefore, efficiently. Hence, one might call it a generalized (or an extended) Granger causality test.[1] This method has not been used so far in the investment-exports-growth context, which this chapter will be concerned with.

A new methodological device of this chapter applies the results proposed by Wickens (1996), and Pesaran and Shin (2002), which imply that for the identification of cointegration coefficients, one needs *a priori*, economic theory-based restrictions on the coefficients. As is now well recognized, the (vector of) coefficients cannot be determined uniquely through the statistical analysis only, hence, we examine a few (exhaustive) economically plausible sets of coefficients, and then adopt a representative set that generates the most general causality results.

This chapter deals with 13 (mainly) developing countries and examines, for each country, whether the postwar growth process was promoted primarily by exports or by domestic investment. One cannot regard Korea as a developing country, but it is included here for the purpose of comparison.

Section 2 describes the methodology of this chapter. Section 3 examines, for each of the 13 countries, the causality relationships mentioned above. Section 4 concludes the chapter.

9.2   VAR Systems, Cointegration, and Causality

The VAR systems to be used in the following include four variables. They are real GDP (*lqp*), real (domestic) investment (*lcp*), real exports (*lep*), and a measure representing financial development, the last of which is either real money stock (real *m2*, *lmp*) or the money multiplier (that is, *m2*/base money, *ln*; King and Plosser (1984) refer to the output expanding effect of the money multiplier or of inside money). Money supply *m2* is broadly defined, including time deposits. Government investment is included in *lcp*. All the variables are expressed in logarithms, and all except for *ln* are measured in per capita terms (that is, they are divided by population size). The variables used in this chapter are described in a List of Symbols following Section 4.

The 13 countries we will focus on are: South Korea (abbreviated to Korea in the following), Colombia, Costa Rica, Chile, the Philippines, South Africa, Turkey, India, Mexico, Malaysia, Pakistan, Venezuela, and Thailand. South Korea is certainly not among developing countries, but is included here for the purpose of comparison. We chose these countries because, while their per capita incomes are largely lower than some OECD members, they are not in the very low-income group.[2] Such countries often have problems regarding data accuracy (see Summers and Heston (1988), who discuss the relationship between data quality and per capita income size). We also exclude countries, such as Argentina and Brazil, which went through very high inflation during (part of) the estimation period.

All the data, which have annual frequency, are drawn from the International Monetary Fund (1979, 1993, 2003, and 2008). The periods used are mostly from 1950 through 2007, but for some countries they are shorter than this maximum length due to lack of data. For each country the period used is shown with estimation results in Table 9.1.

We use trace statistics to decide the number of cointegrating relations (see MacKinnon, Haug, and Michelis (1999); these statistics are known to be more robust than the method using the maximum eigenvalues; see Cheung and Lai (1993)). Appendix A, which is to be omitted from the chapter, shows the trace statistics for the 13 countries (see note 4, Chapter 6 for omitted material). The number of cointegration is one, two, or three. We order the four variables as [*lqp lcp lep lfd*], where *lfd*, standing for a financial development indicator in the logarithm, represents either *lmp* or *ln*.

According to Pesaran and Shin (2002, pp. 54–6), in order that the cointegrated relations can be identified, r^2 restrictions must be distributed across the *r* different cointegrated vectors such that there are *r* restrictions for each of the *r* cointegrated relations. In the two cointegration case with four variables, each relation (each row of cointegrating vectors) has $n - r = 2$ unrestricted variables, where $n(= 4)$ is

Table 9.1 Cointegrated relations and adjustment coefficients

Korea 1961–2007; $obs = 47$; ln; $l = 7$; $co = 2$.

$\alpha_{11} = -0.118$, $\alpha_{21} = 1.180$; $\alpha_{12} = -0.242$, $\alpha_{22} = -0.974$.
(1.533)   (1.607)   (1.732)   (2.586)

$plm = 0.821$, $pwh = 0.618$.

Colombia 1957–2007; $obs = 51$; ln; $l = 6$, $co = 2$.

First cointegration: $lqp = 0.940lcp + 1.054lep - 0.025t$.
         (5.567)   (5.785)   (4.642)

$\alpha_{11} = -0.185$, $\alpha_{21} = -1.470$; $\alpha_{12} = -0.341$, $\alpha_{22} = -2.542$.
(1.862)   (4.396)   (1.756)   (3.900)

$plm = 0.480$, $pwh = 0.689$.

Costa Rica 1957–2007; $obs = 51$; ln; $l = 6$; $co = 2$.

$\alpha_{11} = -0.734$, $\alpha_{21} = 0.862$; $\alpha_{12} = -0.034$, $\alpha_{22} = -0.130$.
(1.472)   (0.570)   (1.594)   (1.985)

$plm = 0.209$, $pwh = 0.292$.

Chile 1968–2007; $obs = 40$; lmp; $l = 4$; $co = 2$.

First cointegration: $lqp = 0.646lcp + 1.136lep - 0.058t$.
         (8.388)   (6.066)   (6.044)

$\alpha_{11} = -0.413$, $\alpha_{21} = -1.044$; $\alpha_{12} = -0.196$, $\alpha_{22} = -0.679$.
(3.525)   (2.946)   (2.600)   (3.058)

$plm = 0.637$, $pwh = 0.403$.

Philippines 1957–2005; $obs = 49$; ln; $l = 6$; $co = 2$.

$\alpha_{11} = 0.370,$ $\alpha_{21} = 1.520;$ $\alpha_{12} = -0.155,$ $\alpha_{22} = -0.449.$
$\quad (3.078) \qquad\qquad (3.196) \qquad\qquad (2.605) \qquad\qquad (1.912)$

$plm = 0.846,$ $pwh = 0.360.$

South Africa 1957–2007; $obs = 51$; lmp; $l = 6$; $co = 3$.

$\alpha_{11} = 0.038,$ $\alpha_{21} = 1.802,$ $\alpha_{31} = -1.899;$ $\alpha_{12} = -0.030,$ $\alpha_{22} = -0.221,$ $\alpha_{32} = -0.098;$
$\quad (0.155) \qquad\quad (2.733) \qquad\quad (1.933) \qquad\quad (1.365) \qquad\quad (3.743) \qquad\quad (1.112)$

$\alpha_{13} = 0.142,$ $\alpha_{23} = 0.362,$ $\alpha_{33} = -0.318.$ $plm = 0.715,$ $pwh = 0.325.$
$\quad (3.024) \qquad\quad (2.893) \qquad\quad (1.706)$

Turkey 1959–2007; $obs = 49$; ln; $l = 8$; $co = 3$.

$\alpha_{11} = 4.515,$ $\alpha_{21} = -3.665,$ $\alpha_{31} = 5.041;$ $\alpha_{12} = -3.710,$ $\alpha_{22} = 1.833,$ $\alpha_{32} = -3.478;$
$\quad (4.181) \qquad\quad (0.726) \qquad\quad (2.114) \qquad\quad (4.093) \qquad\quad (0.433) \qquad\quad (1.732)$

$\alpha_{13} = -1.403,$ $\alpha_{23} = -0.385,$ $\alpha_{33} = -1.976.$ $plm = 0.794,$ $pwh = 0.338.$
$\quad (4.034) \qquad\quad (0.237) \qquad\quad (2.572)$

India 1965–2007; $obs = 43$; lmp; $l = 4$; $co = 3$.

$\alpha_{11} = 0.035,$ $\alpha_{21} = 0.742,$ $\alpha_{31} = -0.542;$ $\alpha_{12} = 0.051,$ $\alpha_{22} = 0.295,$ $\alpha_{32} = -0.335;$
$\quad (0.153) \qquad\quad (2.176) \qquad\quad (0.890) \qquad\quad (0.732) \qquad\quad (2.837) \qquad\quad (1.801)$

$\alpha_{13} = -0.073,$ $\alpha_{23} = -0.038,$ $\alpha_{33} = -0.345.$ $plm = 0.900,$ $pwh = 0.453.$
$\quad (1.603) \qquad\quad (0.559) \qquad\quad (2.829)$

(continued)

Table 9.1 (continued)

Mexico 1956–2007; $obs = 52$; lmp; $l = 5$; $co = 3$.

$\alpha_{11} = -0.153$, $\quad \alpha_{21} = 0.726$, $\quad \alpha_{31} = 0.763$; $\quad \alpha_{12} = -0.028$, $\quad \alpha_{22} = -0.093$, $\quad \alpha_{32} = 0.055$;
$\quad\;\,(0.687)$ $\qquad\;(1.004)$ $\qquad\;\;(0.950)$ $\qquad\quad(1.892)$ $\qquad\quad(1.950)$ $\qquad\quad(1.034)$

$\alpha_{13} = -0.088$, $\quad \alpha_{23} = -0.168$, $\quad \alpha_{33} = -0.386$. $\quad plm = 0.173$, $\quad pwh = 0.464$.
$\quad\;\,(1.941)$ $\qquad\;(1.146)$ $\qquad\;\;(2.370)$

Malaysia 1961–2007; $obs = 47$; lmp; $l = 5$; $co = 1$.

$\alpha_{11} = -0.883$. $\quad lqp = 0.088lcp + 0.063lep + 0.217lmp + 0.013t$.
$\quad\;\,(1.763)$ $\qquad\quad(4.921)\;\;\;\;(3.007)\;\;\;\;\;(5.110)\;\;\;\;\;\;(4.562)$

$plm = 0.029$, $\quad pwh = 0.475$.

Pakistan 1965–2007; $obs = 43$; ln; $l = 4$; $co = 1$; no trend.

$\alpha_{11} = -0.073$. $\quad lqp = -1.229lcp + 1.066lep + 0.892ln$.
$\quad\;\,(3.749)$ $\qquad\quad(2.725)\;\;\;\;\;\;(3.961)\;\;\;\;\;\;(1.981)$

$plm = 0.075$, $\quad pwh = 0.575$.

Venezuela 1958–2007; $obs = 50$; ln; $l = 7$; $co = 2$.

First cointegration: $lqp = 4.465lcp + 0.751lep + 0.513t$.
$\qquad\qquad\qquad\;\;(136.423)\qquad(1.441)\qquad(4.619)$

$\alpha_{11} = -0.765$, $\quad \alpha_{21} = -2.180$; $\quad \alpha_{12} = -3.373$, $\quad \alpha_{22} = -9.633$.
$\quad\;\,(2.169)$ $\qquad\;(1.765)$ $\qquad\quad(2.188)$ $\qquad\quad(1.786)$

$plm = 0.106$, $\quad pwh = 0.467$.

Thailand 1957–2007; $obs = 51$; lmp; $l = 4$; $co = 2$.

First cointegration: $lqp = 0.163lcp + 0.271lep + 0.014t$.
$$\underset{(8.439)}{} \quad \underset{(15.287)}{} \quad \underset{(8.431)}{}$$

$$\alpha_{11} = -1.242, \quad \alpha_{21} = -5.213; \quad \alpha_{12} = -0.076, \quad \alpha_{22} = -0.296.$$
$$\underset{(3.637)}{} \quad \underset{(5.403)}{} \quad \underset{(3.467)}{} \quad \underset{(4.793)}{}$$

$plm = 0.792, \quad pwh = 0.339.$

Notes: (1) Periods after the country names are adjusted periods of estimation. (2) *obs*: the number of observation after adjustment. (3) *lmp* or *ln* is a financial variable for which causalities are detected. (4) Figures in parentheses are the absolute values of *t*-ratios. (5) In each cointegration, a constant is omitted (but appears in the estimation). (6) *t*: a time trend (number of years) with $t = 1$ at 1950. All the cointegrations have a trend term except for Pakistan. Whether the relation has a trend or not is decided by the maximum likelihood, AIC, and BIC; if at least two criteria indicate that the relation should have a trend, a trend is attached. (7) *l*: lag order (lag number). (8) *co*: the number of cointegration. (9) α_{11}, α_{21}, α_{31}: loading factors (adjustment coefficients) relating to the first cointegration; α_{12}, α_{22}, α_{32}: loading factors relating to the second cointegration, etc. (10) Coefficients in italics are relevant to causality detection. (11) *plm*: a *p*-value in the Lagrange multiplier test for serial correlation, with the null of no serial correlation. (12) *pwh*: a *p*-value in the White heteroskedasticity test (no cross terms), with the null of no heteroskedasticity.

the number of variables, $r(=2)$ is the number of cointegrated relations (=the rank of the cointegration matrix) and $r^2(=4)$ is the total number of restrictions to be distributed evenly to the two cointegrated relations (see Wickens (1996), and Pesaran and Shin (2002)). They maintain that cointegrated relations cannot be identified by statistical arguments only, so that the identification requires *a priori*, economic theory-based restrictions on the cointegration coefficients.

When the trace statistics indicate that the number of cointegration is two, we set up the error-correction mechanisms for financial variable *lmp* as

$$
\begin{bmatrix} \Delta lqp_t \\ \Delta lcp_t \\ \Delta lep_t \\ \Delta lmp_t \end{bmatrix} = \begin{bmatrix} \alpha_{11} & \alpha_{12} \\ \alpha_{21} & \alpha_{22} \\ \alpha_{31} & \alpha_{32} \\ \alpha_{41} & \alpha_{42} \end{bmatrix} \begin{bmatrix} 1 & -\beta_{12} & -\beta_{13} & 0 \\ 0 & 1 & -\beta_{23} & -\beta_{24} \end{bmatrix} \begin{bmatrix} lqp_{t-1} \\ lcp_{t-1} \\ lep_{t-1} \\ lmp_{t-1} \end{bmatrix},
$$

where α_{11}, α_{21}, α_{31}, and α_{41} are loading factors (adjustment coefficients) relating to the first cointegration, and α_{12}, α_{22}, α_{32}, and α_{42} are those relating to the second cointegration. The above equations imply that the cointegrated relations, which are shown in the middle matrix (cointegration matrix) on the right-hand side, are

$$lqp_{t-1} = \beta_{12}lcp_{t-1} + \beta_{13}lep_{t-1} \tag{9.1}$$

and

$$lcp_{t-1} = \beta_{23}lep_{t-1} + \beta_{24}lmp_{t-1}. \tag{9.2}$$

In the above relations a constant and a trend term are omitted, but a constant is always included, and, when necessary, a trend term is included in the estimation (see Note 6 to Table 9.1).

Generally, one cannot decide the two cointegrating vectors uniquely (see Enders (2010), pp. 397–8). However, for the seven countries where there appear two cointegrations (Korea, Colombia, Costa Rica, Chile, the Philippines, Venezuela, and Thailand), the alternative, economically plausible cointegration matrix $[1 \ -\beta_{12} \ -\beta_{13} \ 0; \ -\beta_{21} \ 1 \ 0 \ -\beta_{24}]$, where the first four elements are the first row of the 2×4 matrix, yields the same causality results as those in the text. The alternative matrix has a different second row from the original one, the former implying that investment is influenced by per capita output and financial development. The above two cointegration matrices exhaust all the economically plausible ones.

The first cointegrated relation (9.1) implies that, in the long-run, per capita income is raised by increases in per capita capital formation and exports. Note that capital formation and exports, which potentially influence income growth positively, are dealt with on an 'equal footing'. The second relation (9.2) means that capital formation will be affected favorably by increases in exports and financial development. A relationship which assumes exports as a function of investment and financial development is another possibility (here exports come as a second variable before investment), but the above cointegration, investment as a function of exports and so on, is generally a more plausible one (because a big factor that influences exports is economic conditions of foreign countries, and so on). In a two cointegrated system, the minimum number of restrictions (the form of restrictions being 1 or 0) is $r^2 = 4$, where r is the number of cointegration or the rank of the cointegration matrix; an extra $\beta_{ij} = 0$ in the second or third column makes the system still identifiable, but here we choose a more general setting of minimum restrictions. An extra $\beta_{ij} = 0$ is called an over-identifying restriction (see Pesaran and Shin (2002)).

In the case of two cointegrated relations, the first two VAR equations with error-correction terms ($\alpha_{ij}[CRj], j = 1, 2$) can be written as

$$\Delta lqp_t = a_1 + \Sigma a_{1k}\Delta lqp_{-k} + \Sigma a_{2k}\Delta lcp_{-k} + \Sigma a_{3k}\Delta lep_{-k} + \Sigma a_{4k}\Delta lfd_{-k}$$
$$+ \alpha_{11}(CR1) + \alpha_{12}(CR2),$$

where $CR1 \equiv \delta_1 + lqp_{t-1} - \beta_{13}lcp_{t-1} + \gamma_1 t, CR2 \equiv \delta_2 + lcp_{t-1} - \beta_{23}lep_{t-1} - \beta_{24}lfd_{t-1} + \gamma_2 t$, and

$$\Delta lep_t = a_2 + \Sigma b_{1k}\Delta lqp_{-k} + \Sigma b_{2k}\Delta lep_{-k} + \Sigma b_{3k}\Delta lcp_{-k} + \Sigma b_{4k}\Delta lfd_{-k}$$
$$+ \alpha_{21}(CR1) + \alpha_{22}(CR2),$$

(Other equations with Δlcp and Δlfd as a dependent variable are omitted.) Here, k runs from 1 to an appropriate number which is decided by the requirements that the VAR system shows some causality from exports to output or from investment to output, as well as that the VAR system passes the Lagrange multiplier tests (LM tests) for serial correlation; the lag numbers adopted in the following coincide in most of the cases with the optimal numbers indicated by the AIC (Akaike Information Criterion). In the above system, a_i, a_{ij}, b_{ij}, α_{ij}, β_{ij}, γ_i, and δ_i are all constants to be estimated, and t is a trend term (a number of years). The general VAR system, when the two cointegrating vectors are involved, can be shown in a similar way.

When the trace statistics indicate that the number of cointegrated relations is three, we set up the cointegrated matrix as

$$
\begin{bmatrix}
1 & -\beta_{12} & 0 & 0 \\
0 & 1 & 0 & -\beta_{24} \\
-\beta_{31} & 0 & 1 & 0
\end{bmatrix}
\tag{9.3}
$$

where the four variables are ordered as [*lqp lcp lep lfd*], and *lfd* is one of the two financial variables. In (9.3), as in the two cointegrated vector case, $r^2 = 9$ restrictions are distributed evenly across the three cointegrated vectors, with three restrictions on each vector (one being a normalizing restriction).

The first cointegrated vector in (9.3) implies the investment-multiplier relationship. The second vector means that investment is prompted by the development of financial systems. The third relation says that per capita exports are promoted by larger per capita income. Note that $r^2 = 9$, where r is the number of restrictions (number of terms other than β_{ij}) and also the number of the rank of the above matrix. Also, an extra β_{ij} makes the above three cointegration system unidentifiable.

In the four countries that have three cointegrations (South Africa, Turkey, India, and Mexico), the alternative, economically plausible cointegration matrix [1 0 $-\beta_{13}$ 0; 0 1 0 $-\beta_{24}$; 0 $-\beta_{32}$ 1 0], where the first four elements are the first row of the 3×4 matrix, yields the same causalities, or the latter matrix generates causalities that are included in the results from the matrix (9.3). (In this alternative matrix, output is supposed to respond more to exports than to investment; in this matrix, the third row cannot be [$-\beta_{31}$ 0 1 0], because this causation is not compatible with the first row's.) The above consideration exhausts all the economically plausible matrices when the cointegration number is three. (If the second row is [0 1 $-\beta_{23}$ 0], the rank of the matrix becomes two.)

For the variables in the VARs to form cointegration(s), at least two of them have to be unit-root variables. To test this property we use the KPSS test (see Kwiatkowski et al. (1992)), which is known to be more powerful in rejecting the null of the unit-root hypothesis than other conventional tests.[3]

9.3   Investment, exports, and economic growth

In this section we conduct the causality tests for the 13 countries. First, in Korea, financial variable *ln* yields two cointegrations with *l* (lag order) = 7; see Table 9.1. For the first (resp. second) cointegrating vector to be meaningful, the loading factor α_{11} (resp. α_{22}) should be negative and significant (see Luintel and Kahn (1999)). But the eligible cointegration is only the second one; see Equation (9.2), so that this cointegrated relation does not indicate the (long-run) causality we are looking for. But the adjustment coefficient $\alpha_{12} = -0.242$ (1.732, which is a *t*-ratio in absolute value; the corresponding *p*-value in a the one-sided test is 0.045; one-sided test is applied here because the sign of α_{12} is not relevant) indicates that there is a causality from investment to per capita income growth.

Lag order $l < 7$ did not yield any causality; neither did financial variable *lmp* for any lag order. Note that investment (resp. per capita income) is a second (resp. first) variable in the VAR and is represented by a second (resp. first) suffix of α_{12}. See Table 9.1 for the estimation period and observation number, as well as other detailed information on Korea. In particular, the *p*-values in the Lagrange multiplier test and White heteroskedasticity test imply that the residuals are white noise, so that the second long-run relation (not shown in the table because it does not tell any causality one is interested in) satisfies the condition for cointegration (see Notes 11 and 12 to Table 9.1).

In the case of Colombia, we found two cointegrated relations with negative and significant loading factors for *ln* and *l* = 6. The first cointegration, which is listed in the table, shows that there are long-run effects from investment and exports to per capita output, which, considering the two *t*-ratios as well, have about the same size of influence. A significant α_{12} also implies that there is a causality from investment to per capita output growth.

We next turn to Costa Rica, which shows one meaningful cointegration with investment as a dependent variable. It, therefore, does not give any causality that we are interested in. The financial variable is *ln*, and *l* = 6. Also, α_{12}'s one-sided *t*-ratio is 1.594, with $p = 0.059$, so that this coefficient is significant only at the 6% level. However, since other clues to any causality are not available, we adopt α_{12}'s *t*-ratio to conclude that there is a causality from investment to per capita output growth.

Chile's postwar period yields two meaningful cointegrations for *lmp* and $l = 4$. The first one, Equation (9.1), which is relevant to causality detection, is shown in the table. There are causalities from investment and exports to per capita output. Note that the effect of exports is somewhat stronger than that of investment, because the ratio between the coefficients is $1.136/0.646 = 1.759$. A significant α_{12} also implies that a causality runs from investment to per capita output.

The Philippines yields two cointegrations for *ln* and $l = 6$, but a negative and significant loading factor is found only with the second, which does not show a causality from one of the two demand factors to output. However, since α_{12} is significant ($t = 2.605$), one can detect a causality from investment to per capita output.

South Africa shows three cointegrated relations according to trace statistics, and the second and third relations have negative and significant loading factors. The financial variable is *lmp*, and $l = 6$. Further, a significant α_{13} implies that exports caused per capita output growth.

Postwar Turkey's experience shows an interesting but puzzling result. First, the VAR order [*lqp lcp lep lfd*] did not yield any causality for any lag order. Hence we try VARs where the order of *lcp* and *lep* is switched, letting *lep* come before *lcp*. In the second row of a new 3×4 matrix, exports (not investment) are supposed to be affected by financial development, which is less plausible than our first 3×4 matrix. Then as the table exhibits, for *ln* and $l = 8$, three cointegrations are possible, and the last one is meaningful. Also, one finds that α_{13} is significant ($t = 4.034$), showing that investment caused output growth (note that the third variable in this case is investment, not exports); for financial variable *lmp*, $l = 7$ yielded the same result, with the same order of variables, but since the t-ratio of α_{13} is 1.709, Table 9.1 exhibits the first case.

Although India's period of observation is rather shorter, it yielded one meaningful cointegration out of three possible relationships, for *lmp* and $l = 6$. In the third cointegration, the t-ratio of $\alpha_{13} = 1.603$ ($p = 0.058$), so that the coefficient is significant only at the 6% level; however, this is the most significant one for this country, hence we adopt it, which implies that there is a causality from exports to output growth.

In Mexico's case, trace statistics say that there are three cointegrations and the second and third are meaningful, when the financial variable is *lmp*, and $l = 5$. Significant α_{12} and α_{13} imply that causalities run from both investment and exports, but the size of coefficients indicates that the effect of exports is much stronger than that of investment; $0.088/0.028 = 3.143$.

We next turn to Malaysia. For *lmp* and $l = 5$, only one meaningful cointegration was found, which is shown in Table 9.1. The coefficients on *lcp* and *lep* are significant and nearly of the same size. But the *p*-value in the Lagrange multiplier test for $l = 5$, with the null of no serial correlation, is 0.029, implying that there is a serial correlation for $l = 5$ (but not for $l < 5$).

Hence we tried the conventional Granger causality tests (*F*-tests), which support the above results, implying that investment and exports are both causal factors on per capita output. Let the sum of squared residual (SSR) from the OLS of *lqp* on *lcp*(−1), *lep*(−1), *lmp*(−1), a constant, and a trend be S_2. Let the SSR from the OLS when *lcp* (−1) (*lep*(−1)) is dropped from the right-hand side be S_0 (S_1). Then one finds $S_2 = 0.051$, $S_0 = 0.065$, and $S_1 = 0.120$. The *F* statistic regarding *lcp* is $[(120 − 51)/1]/[51/(47 − 5)] = 56.82$, where the adjusted observation number is 47, explanatory variables (including a constant and a trend) are five, and the constraint number is one. Similarly, for *lep*, $F = [(65 − 51)/1]/(47 − 5) = 11.53$.

The upper 1% (5%) point of the *F*-distribution with $df\,(1,42)$ is 8.80 (4.07), hence the null that *lcp*'s (*lep*'s) coefficient is zero can be rejected at the 1% level.

For Pakistan, one obtains, for *ln* and $l = 7$, one cointegration that is meaningful. Although the sign of *lcp*'s coefficient is opposite to what is expected ('output and investment are substitutes' or 'investment crowds out output'), the *lep*'s coefficient is positive and significant, hence we conclude that exports are the causal factor in postwar Pakistan's economic growth. Although the *p*-value in the Lagrange multiplier test is 0.075, meaning that the null of no serial correlation cannot be rejected only at the 8% level, the corresponding traditional Granger tests, with one lag in explanatory variables and a trend, which are omitted here, support the above result.

Turning to Venezuela, we find two meaningful cointegrations for *ln* and $l = 7$. The first cointegration exhibits that only the *lcp*'s coefficient is significant. Also, α_{12} is significant. Hence they jointly imply that investment caused output growth.

Table 9.2 A summary of causal demand factors

Korea: $lqp \leftarrow lcp$.	**Colombia:** $lqp \leftarrow lcp, lep$.	**Costa Rica:** $lqp \leftarrow lcp$.
Chile: $lqp \leftarrow lcp, lep^*$.	**Philippines:** $lqp \leftarrow lcp$.	**South Africa:** $lqp \leftarrow lep$.
Turkey: $lqp \leftarrow lcp$.	**India:** $lqp \leftarrow lep$.	**Mexico:** $lqp \leftarrow lcp, lep^*$.
Malaysia: $lqp \leftarrow lcp, lep$.	**Pakistan:** $lqp \leftarrow lep$.	**Venezuela:** $lqp \leftarrow lcp$.
Thailand: $lqp \leftarrow lcp, lep^*$.		

Notes: (1) $\leftarrow$: a direction of causality. (2) A variable with * has a stronger causality.

We now turn to the last country, Thailand. It yields two meaningful cointegrations for *lmp* and $l = 4$. The first cointegration shows that both investment and exports caused Thailand's economic growth. But the two coefficients indicate that the influence of exports was slightly larger in raising Thailand's per capita income; $0.271/0.163 = 1.663$.

9.4 Conclusions

A new methodological device of this chapter has been that, applying the statistical analyses of Wickens (1996) and Pesaran and Shin (2002), it was possible to set up economically plausible and statistically significant sets of cointegrated relationships and choose a set that exhibits the most comprehensive causality results, when the number of cointegrations is either two or three.

It will be convenient to summarize the results so far obtained, which are listed in Table 9.2. Of the 13 countries sampled, five countries have both investment and exports as causal factors on output growth. Five other countries were influenced mainly by investment in output growth, while the remaining three countries were affected by exports in their output growth. Among the first five countries, three countries (Chile, Mexico, and Thailand) have exports which are stronger in their influence than investment, hence one may be able to say that the effects of investment and exports are almost evenly distributed among the countries examined.

It would also be of some interest to note that the countries which show the causality results when the financial variable is *lmp* (the per capita real money stock) are six, while the countries showing causalities with financial variable *ln* (the money multiplier) are seven, so that the relevant financial variables are also nearly equal in number among the sampled countries.

List of Symbols

lqp   real per capita GDP, where the initial *l* represents that it is expressed in the natural logarithm, which also applies to other symbols.

lcp   real per capita (domestic) investment, including that by the government.

lep   real per capita exports.

lmp   real per capita *m2*, which includes time deposits.

ln   *m2*/base money, that is, the money multiplier regarding *m2*.

lfd   financial development indicator, which is either *lmp* or *ln*.

α_{ij} loading factor (a speed of adjustment coefficient) in an error-correction mechanism.

β_{ij} coefficient in a cointegrated relation.

l   lag order (lag number) of a VAR system.

Notes

1. The above motivation (comparison of investment and exports as a stronger growth-promoting factor) would be backed up from another angle. In the analytical descriptions of Japanese 'high-growth period' which lasted from 1955 (or 1950) through 1973, vigorous domestic investment and exports were always mentioned as Japan's growth boosters; see, for example, Takenaka (1991, chs 1, 5, and 6). However, in the Japanese case also, the problem of which demand factor, investment or exports, were stronger in leading its postwar high growth has not been addressed in any statistical or econometric way.
2. Korea, Turkey, and Mexico are, however, OECD members.
3. The test can show that for 11 countries, excluding South Africa and Venezuela, at least two member variables, which are used in causality detection, have a unit root. But in the above two countries, one can show only the same results using the augmented Dickey-Fuller test. The test results (Appendix B) are omitted from the paper but are available from the author upon request.

   All the computations reported in this chapter were done with EViews 6 (Quantitative Microsoftware, 2007).

References

Arestis, P. and P.O. Demetriades (1997) Financial development and economic growth: assessing the evidence. *Economic Journal* 107, 783–99.

Arestis, P., P.O. Demetriades, and K.B. Luintel (2001) Financial development and economic growth: the role of stock markets. *Journal of Money, Credit, and Banking* 33, 16–41.

Armad, J. and S. Harnhirun (1995) Unit roots and cointegration in estimating causality between exports and economic growth: empirical evidence from the ASEAN countries. *Economics Letters* 49, 329–34.

Barro, R.J. and X. Sala-i-Martin (2004) *Economic growth*, 2nd ed. Cambridge, Mass.: MIT Press.

Cheung, Y.-W. and K.S. Lai (1993) Finite-sample sizes of Johansen's likelihood ratio tests for cointegration. *Oxford Bulletin of Economics and Statistics* 55, 313–28.

DeLong, J.B. and L.H. Summers (1991) Equipment investment and economic growth. *Quarterly Journal of Economics* 106, 445–502.

Demirguc-Kunt, A. and Levine, R., eds. (2001) *Financial structure and economic growth: a cross-country comparison of banks, markets, and development.* Cambridge, Mass.: MIT Press.

Devereux, M.B. (1997) Growth, specialization, and trade liberalization. *International Economic Review* 38, 565–85.

Enders, W. (2010) *Applied Econometric Time Series*, 3rd ed. New York: Wiley.

Feder, G. (1982) On exports and economic growth. *Journal of Development Economics* 12, 59–72.

Hall, S.G. and A. Milne (1994) The relevance of P-star analysis to UK monetary policy. *Economic Journal* 104, 597–604.

International Monetary Fund (1979, 1993, 2003, 2008) *International Financial Statistics*, Washington, D.C.

King, R.G. and C.I. Plosser (1984) Money, credit, and prices in a real business cycle. *American Economic Review* 74, 363–80.

Krueger, A.O. (1997) Trade policy and economic development: how we learn. *American Economic Review* 87, 1–22.

Kwiatkowski, D., P. Phillips, P. Schmidt, and Y. Shin, (1992). Testing the null hypothesis of stationarity against the alternative of a unit root. *Journal of Econometrics* 54, 159–78.

Levine, R. and D. Renelt (1992) A sensitivity analysis of cross-country growth regressions. *American Economic Review* 82, 942–63.

Luintel, K.B. and M. Khan (1999) A quantitative reassessment of finance-growth nexus: evidence from a multivariate VAR. *Journal of Development Economics* 60, 381–405.

MacKinnon, J.G., A.A. Haug, and L. Michelis (1999). Numerical distribution functions of likelihood ratio tests for cointegration. *Journal of Applied Econometrics* 14, 563–77.

Madsen, J.B. (2002) The causality between investment and economic growth. *Economics Letters* 74, 157–63.

Pesaran, M. and Y. Shin (2002) Long-run structural modeling. *Econometric Reviews* 21, 49–87.

Riviera-Batiz, L. and P.M. Romer (1991) Economic integration and endogenous Growth. *Quarterly Journal of Economics* 56, 531–55.

Summers, R. and A. Heston (1988) A new test of international comparisons of real product and price levels: estimates for 130 countries, 1950–1985. *Review of Income and Wealth* 36, 1–26.

Takenaka, H. (1991) *Contemporary Japanese Economy and Economic Policy*. Ann Arbor: University of Michigan Press.

Wickens, M.R. (1996) Interpreting cointegrating vectors and common stochastic trends. *Journal of Econometrics* 74, 255–71.

Yamada, H. (1998) A note on the causality between export and productivity: an empirical re-examination. *Economics Letters* 61, 111–4.

The following Appendices A and B, as well as the data for this chapter are omitted, but are available from the author (see Note 4, Chapter 6).

Appendix A. Trace Statistics of the 13 Countries

Appendix B. Unit Root Tests (KPSS Test Statistics)

Names Index

Subject Index